BLOOMS

Con

C

Bird

Guide

There are 47 individual Wildlife Trusts covering the whole of the UK and the Isle of Man and Alderney. Together The Wildlife Trusts are the largest UK voluntary organization dedicated to protecting wildlife and wild places everywhere – at land and sea. They are supported by 800,000 members, 150,000 of whom belong to their junior branch, Wildlife Watch. Every year The Wildlife Trusts work with thousands of schools, and their nature reserves and visitor centres receive millions of visitors.

The Wildlife Trusts work in partnership with hundreds of landowners and businesses in the UK. Building on their existing network of 2,200 nature reserves, The Wildlife Trusts' recovery plan for the UK's wildlife and fragmented habitats, known as A Living Landscape, is being achieved through restoring, recreating and reconnecting large areas of wildlife habitat.

The Wildlife Trusts also have a vision for the UK's seas and sea life – Living Seas, in which wildlife thrives from the depths of the oceans to the coastal shallows. In Living Seas, wildlife and habitats are recovering, the natural environment is adapting well to a changing climate, and people are inspired by marine wildlife and value the sea for the many ways in which it supports our quality of life. As well as protecting wildlife, these projects help to safeguard the ecosystems we depend on for services like clean air and water.

All 47 Wildlife Trusts are members of the Royal Society of Wildlife Trusts (Registered charity number 207238). To find your local Wildlife Trust visit wildlifetrusts.org

BLOOMSBURY

Concise Coastal Bird Guide

BLOOMSBURY

LONDON • NEW DELHI • NEW YORK • SYDNEY

Bloomsbury Natural History
An imprint of Bloomsbury Publishing Plc

50 Bedford Square
London
WC1B 3DP
UK

1385 Broadway
New York
NY 10018
USA

www.bloomsbury.com

First published 2015

A catalogue record for this book is available from the British Library.

Library of Congress Cataloguing-in-Publication data has been applied for.

ISBN: PB: 978-1-4729-2179-6
ePub: 978-1-4729-2180-2

4 6 8 10 9 7 5 3

Printed and bound in China by Leo Paper Group

Contents

Introduction 6

Wildfowl 12
Divers 42
Grebes 45
Fulmar and Shearwaters 50
Storm-petrels 56
Gannet 58
Cormorants 59
Herons 61
Spoonbills 65
Hawks & Allies 66
Osprey 71
Falcons 72
Rails 76
Oystercatchers 80
Stilts 81
Avocets 82
Plovers 83
Waders 88
Skuas 112
Gulls 116
Terns 130
Auks 136
Pigeons 141
Cuckoos 142
Barn Owls 143
Typical Owls 144
Swifts 146
Kingfishers 147
Woodpeckers 148
Larks 150
Swallows & Martins 152
Pipits & Wagtails 155
Chats 161
Thrushes 165
Warblers & Allies 166
Flycatchers 174
Bearded Tit 175
Crows 176
Starlings 181
Finches 182
Buntings 185

Index 190

Introduction

Birds are the most diverse and the most visible of our larger land animals and on the coastline, birdwatching is particularly rewarding, partly because large numbers of birds congregate at certain kinds of coastal habitats, and partly because the open landscape of coastal environments makes observation easier.

Birds of coastal habitats

There are several bird families which are rarely seen anywhere except coasts and at sea. They include auks, terns and shearwaters – birds that breed on cliffs, clifftops or beaches and spend their winters ranging many miles offshore. Other bird groups have strong ties to the coast, finding much of their food on, in or by the sea at least for part of the year, but also spend some of their time inland – they include gulls, some ducks and geese, and most wading birds. Then there are the birds that have no particular requirement to live or breed by the sea but nevertheless often do use coastal habitats as well as inland ones.

Bird movements are not completely predictable and in theory, any of the 200 or so British breeding birds, and the many others that are either common or rare non-breeding visitors, could be observed in a coastal setting. However, this book limits its scope to the 180 species that are particularly typical of seaside habitats and most likely to be encountered by people exploring the coastlines of Britain.

Habitat types

Where land meets sea, there may be a gently sloping sand or pebble beach, a vast expanse of marshland that grades into saltmarsh and muddy shore, a sheer cliff-face, or a more gradual series of rocky 'steps' to sea level. Depending on the angle of the shoreline relative to prevailing wave direction, some shores become eroded while in other areas the sea deposits material, building beaches. Hard rock erodes more slowly than soft rock, and the type of material carried downriver influences the types of beaches that form near the river's outflow into the sea. Soil composition and exposure to wind influences the kinds of vegetation that will grow on clifftops and above the waterline on beaches.

Different birds are adapted to exploit different kinds of coastal habitat. Many of our 'true' seabirds – those which habitually forage actually on or in the sea, make use of cliff-faces, as these offer great protection from predatory mammals like rats and Foxes. Islands that are entirely free of such predators are of even higher importance to seabirds. Because the seabirds find their food in the open ocean, they have no need to defend a territory on land beyond the immediate surroundings of their nests, so can breed in huge and dense colonies. These 'seabird cities' are one of the most dramatic wildlife spectacles available anywhere in the world – the sensory impact can even be almost overwhelming. By the end of summer, though, such colonies are deserted, their occupants roaming distant seas. These birds have no need to go near land at all for any purpose other than breeding.

Estuaries and mudflats are also of great importance to birds. On saltmarshes, dense beds of eelgrass feed the thousands of Brent Geese that migrate to Britain from the Arctic every winter. The expanses of mud revealed by the outgoing tide are rich with organic material and support huge quantities of burrowing invertebrate animals – tiny gastropod snails, various kinds of worms, bivalve molluscs such as cockles, and crustaceans including shrimp and crabs. These provide food for wading birds including sandpipers, godwits and plovers, which form enormous flocks at key estuarine 'staging posts' during their southbound autumn migration. These flocks may include 10 or more different species, which avoid direct competition by specialising in different prey – the differences in their bill shapes indicates whether they are deep or shallow probers, or surface pickers.

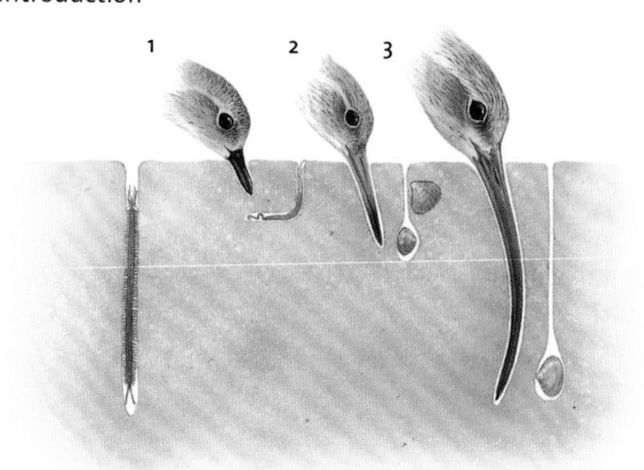

Many wading birds have specially adapted bills that help them extract food from sand or mud. Plovers (1), Redshanks (2) and Curlews (3) all have very different bills as they hunt for food at different depths in the sand.

Beaches and rocky shores do not attract such vast numbers of birds, but support a range of specialised representatives from a range of bird families. Undisturbed sandy and shingly beaches are breeding grounds for terns and Ringed Plovers, and in winter are foraging grounds for Snow Buntings and Shore Larks. Rocky coasts that are rich with mollusc, crustacean and other shoreline life will attract certain waders such as Turnstones and Oystercatchers, and also support the Rock Pipit, our most strictly coastal songbird.

Where reedy marshland and shallow saline lagoons lie close to the sea edge, a rich and diverse community of birdlife will develop. Avocets and Black-tailed Godwits feed in the shallow, sheltered water, joined by an array of dabbling ducks in winter, while the extensive reedbeds surrounding the open water provide nesting and feeding habitat for the likes of Marsh Harrier, Bittern and Bearded Tit. This kind of habitat is also found inland in some low-lying areas with large rivers or lakes, but most of our finest and richest marshlands lie on the coast.

Coastlines of every kind are also the best places to find and watch migrating songbirds. Headlands are particularly good, as a bird crossing the sea will aim for the first bit of land it can see, and if it is tired from its journey it will probably spend some time in the immediate area, resting and looking for food. Headlands serve as 'launchpads' for outgoing migrants that are seeking as short a sea crossing as possible. Additionally, they are great places for sea-watching, as migrating seabirds making their way around a coast tend to cut straight across the bays but are more likely to pass closer to a headland. Weather plays a major role in the movements of migrating seabirds and land birds, with storms at sea pushing seabirds nearer to land, and rain and fog making migrants more likely to stay put.

Our coastline is highly developed, and this inevitably reduces its value for wildlife. Particularly in southern areas, most beaches are too disturbed to support tern colonies, and many estuarine areas have been drained, their waters channelled through concrete, and the recovered land built upon. Pollution from shipping, run-off and littering can ruin otherwise intact habitats, and harm and kill individual birds, while overfishing can devastate seabird populations. Development also interferes with the natural process of coastal reshaping. The losses through the 20th century have been enormous, but as we move through the 21st century there is a fast-growing awareness of the need to safeguard wildlife and habitats. Ambitious projects to restore some coastal habitats are already paying dividends.

The basics of identification

Some birds are very easily recognisable even from a brief or poor view, possessing an outline or plumage pattern that is quite unlike any other species. The Puffin is one example of this, the Oystercatcher another. More often, it is easy to narrow the identification down to a particular group but not so easy to pin down the exact species. Sometimes, checking one specific feature will be enough for species identification – for example, the Sanderling is our only common wading bird that doesn't have a hind toe. But more often, you will need to take into account a variety of features, and assessing these features may be a much more subjective process than counting the number of toes!

When you see a bird, there is a lot of information to take in, sometimes in a short space of time, but not all of that information will be useful. If your bird is on its own and there are no familiar objects around for reference, your estimate of its size may well be inaccurate. Your perception of colour can also be unreliable or unhelpful, as this changes considerably with the light, and in some species shows considerable individual variation. Instead, concentrate on the bird's shape, the relative proportions of its body parts (especially bill, legs and wings), and the pattern of its plumage – looking for areas of contrast. Also take note of the way the bird moves. If it is on the ground, does it hop, run or walk? If flying, does it glide or flap fast, is its flight line straight or bouncy? If on the water, does it dive, and if so how long does it stay under?

Birds, unlike other vertebrate groups, have a remarkably consistent 'body plan', with the same arrangement of feather tracts (groups of similar-sized and -shaped feathers on particular parts of the body), so once you understand the basic layout of a generic bird, you can apply that to virtually every real bird you encounter. Learning names of the major body parts and feather tracts is worthwhile. It helps you to interpret descriptions in field guides and will improve your identification skills. The difference between two similar species can just be a different tone or pattern in one or two feather groups.

Bird songs and calls can help with identification but is not as useful on the coast as it is in woodland – seabirds, for example, are usually silent when feeding. But learning the flight calls of waders can help you pin down a fast-moving shape going overhead, and alert you to the approach of something interesting heading your way.

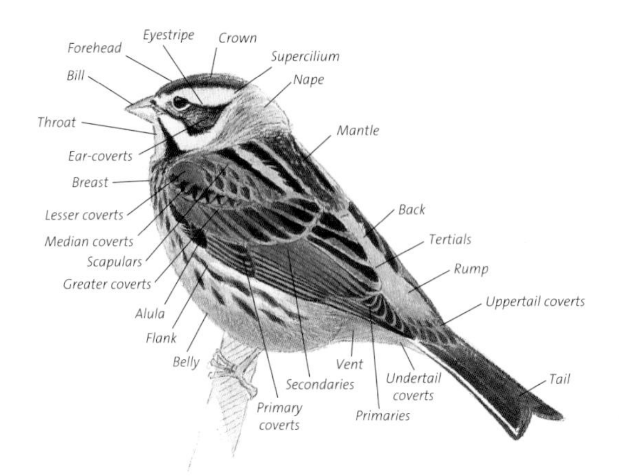

The more common birds you can learn to recognise, the more likely you are to quickly realise when you are looking at (or listening to) something that is out of the ordinary.

Using this book

The accounts in this book have four sections. **SIZE AND DESCRIPTION** gives the bird's length in centimetres and summarises its appearance, noting any key features that will help with identification. **VOICE** describes its main calls and song. **HABITAT** describes the bird's distribution in Britain, whether it is resident (present all year round), a summer or winter visitor only, or a passage migrant (occurs in spring and autumn, on its way to or from breeding and wintering areas elsewhere). It also describes the kinds of environment where you are most likely to see the bird. Finally, **FOOD AND HABITS** explains what the bird eats, how it forages, and any notable information on its general behaviour.

Bewick's Swan

Cygnus columbianus

Size and description 125cm. The smallest of our swans. Adult pure white with black bill, small rounded yellow marking on bill base, eyes and legs black. Juvenile light grey with same bill pattern as adult but marking is pinkish rather than yellow.
Voice High-pitched bugling call, softer than Whooper's.
Habitat Breeds on Russian Arctic tundra, migrates to north-west Europe, including mainly southern Britain, in winter. Damp meadowland and shallow lakes.
Food and habits Feeds on grass, aquatic vegetation and other plant matter. Often stays in family groups within larger flock. Highly gregarious and may flock with other swan species.

Whooper Swan

Cygnus cygnus

Size and description 150cm. About the size of Mute Swan but less bulky. Adult pure white with black bill, large yellow marking on bill base extending in point towards bill tip, eyes and legs black. Juvenile light grey with same bill pattern as adult but marking is pinkish rather than yellow.
Voice Loud, ringing, bugling call.
Habitat Breeds in northern and eastern Europe, visits Britain and western Europe in winter, more common in northern Britain. Damp meadowland and shallow lakes.
Food and habits Diet and habits very similar to those of Bewick's Swan. Often on arable fields, sometimes with Mute Swans.

Bewick's Swan
adult
Bewick's Swan
juvenile
Whooper Swan
juvenile
Whooper Swan
adult

Mute Swan

Cygnus olor

Size and description 152cm. Largest flying bird in Britain. Adult is all white, juvenile grey-brown. Distinguished from wintering Bewick's and Whooper Swans by orange bill with black knob at base (smaller in female) and more graceful curve to neck.

Voice Generally silent; hisses when angry or disturbed.

Habitat Almost any still or slow-moving inland water body; also estuaries and sheltered coastal regions. Found across northern and western Europe. Resident in Britain.

Food and habits Usually feeds on water by dipping its neck below the surface, sometimes up-ending. Nest a large mound of plant matter on the edges of water bodies.

Pink-footed Goose

Anser brachyrhynchus

Size and description 70cm. Small and neat grey goose with a rounded head and short neck that are darker than the rest of its body. The legs are pink, and there is a pink band on the bill.

Voice Vocal; call a ringing 'ung unk' and 'wink wink wink'.

Habitat Breeds in Iceland and Spitsbergen. Winters on coastal meadows in Britain and the Low Countries. In Britain found predominantly in Norfolk, Lancashire and eastern Scotland).

Food and habits Diet is almost entirely vegetarian.

▶ **Similar species Bean Goose** (*A. fabalis*) 75cm long. Orange legs, narrow orange band on bill and darker plumage than that of Pink-footed Goose. Breeds in north-east Europe and Siberia; winters in coastal wetlands in north-west Europe. Scarce in Britain; mainly on east coast.

White-fronted Goose

Anser albifrons

Size and description 70cm. Two races appear in large wintering flocks in Britain. Greenland White-fronts (*A. a. flavirostris*) winter in Ireland and Scotland, smaller Russian birds (*A. a. albifrons*) in Wales and southern England. Greenland birds have orange bills and legs, Russian ones pink bills and orange legs. Adult of both races has a white forehead and black belly markings.

Voice Call higher pitched than that of other grey geese, and with a whinnying quality.

Habitat Breeds on Arctic tundra. Winter visitor to Britain, on rough grassland, saltmarsh, freshwater marsh and farmland.

Food and habits Feeds on herbs, grasses and sedges, as well as agricultural grain, potatoes and sprouting cereals (the latter particularly in winter).

Greylag Goose
Anser anser

Size and description 83cm. Large grey goose with an orange bill and flesh-coloured legs.

Voice Calls in flight 'aahng-ung-ung'.

Habitat Marshy moorland during breeding season. Winter visitor to salt-water and freshwater marshes, grasslands and estuaries. Feral birds present in Britain throughout the year, but wild birds are mostly winter visitors.

Food and habits Diet mostly grass, but also cereals in autumn and winter. Nest a down-lined cup on the ground.

Canada Goose

Branta canadensis

Size and description 97cm. A large grey-brown goose with a black head and neck, and a white patch on the chin that extends up the head. Distinguished from Barnacle and Brent Geese by larger size, longer neck and preference for inland habitats.

Voice Loud trumpet-like call, 'ah-honk'.

Habitat Resident on inland and coastal fresh water of all kinds. Introduced to Britain and now widespread, a few vagrants from North America turn up on north-west coasts.

Food and habits Feeds mainly on grass. Nest a down-lined cup, usually near water.

Brent Goose

Branta bernicla

Size and description 60cm. Smallest British goose, very dark with black head and neck, dusky blackish upperside and grey barred underside. Two subspecies visit Britain, the dark-bellied *bernicla* and the pale-bellied *hrota*; more contrastingly marked North American subspecies 'Black Brant' (*nigricans*) is a rare visitor.

Voice A low, rolling 'rrott'.

Habitat Breeds in northern and eastern Europe, winter visitor to Britain, dark-bellied birds to southern coasts, pale-bellied birds to northern and Irish coasts. Estuaries and saltmarsh, sometimes arable fields.

Food and habits Feeds mainly on eel-grass in the intertidal zone, often swims on sea. Gregarious, flies in straggling clumpy flocks or lines rather than Vs.

Barnacle Goose

Branta leucopsis

Size and description 65cm. Attractively marked, small-billed grey, black and white goose, with white face and forehead, black crown, eyestripe and neck, grey and black barred upperside and lighter barred underside.

Voice A sharp yapping bark.

Habitat Breeds in Arctic Europe, winter visitor to mainly north-west Britain. There are also a few resident feral populations scattered in southern England, and escapees from ornamental wildfowl collections are frequently encountered. Grasslands and shallow lakes.

Food and habits Feeds mainly on grass. Sociable and noisy. Lone birds will flock with other goose species.

Barnacle Goose

Shelduck

Tadorna tadorna

Size and description 61cm. Large duck with bold markings: head and neck dark green, wide chestnut breast band, black on wingtips and end of tail, and white underparts. Sexes are similar.

Voice Generally silent, but drake can give a whistle when in flight. Female quacks.

Habitat Can be seen year-round on estuaries, sandy shores and saltmarsh. Breeds mainly on coasts, occasionally inland on islands or banks of lakes and rivers.

Food and habits Feeds chiefly on small molluscs caught by sweeping bill through soft estuarine mud. Nests in dunes in rabbit burrows.

Wigeon
Anas penelope

SIZE AND DESCRIPTION 18cm. Drake has a chestnut head with a creamy yellow stripe from the bill over the crown, pinkish breast and short black-tipped blue bill. Speculum is green. White patches on the wings are visible in flight.

VOICE Drake has a whistling 'whee-oo' call.

HABITAT Marshes. Breeds in north. Winter visitor to much of central and southern Europe, including Britain, often on coastal marshes and estuaries, but also inland.

FOOD AND HABITS Eats mostly plant matter, which it takes from the water's surface. Often seen in flocks grazing on land.

Gadwall

Anas strepera

Size and description 51cm. Male grey with a blackish rump and bill. Female similar to female Mallard, but smaller. Both sexes have a characteristic white speculum. Legs are orange-yellow.

Voice A wooden 'errp'.

Habitat Mainly inland waters, and occasionally coastal marshes or estuaries. Present in Britain throughout the year, but rarely breeds.

Food and habits Diet consists of seeds, leaves, roots and stems of aquatic plants, as well as grasses and stoneworts; occasionally also cereal grains on land. Not as gregarious as some other dabbling ducks outside breeding season, tending to form only small flocks.

Common Teal

Anas crecca

Female

Male

Female

Male

Size and Description 35cm. Smaller and neater in appearance than Mallard. Drake has a chestnut head with a green eyestripe, speckled breast and creamy undertail. Speculum is green.

Voice Drake gives a whistling 'crrick, crrick' call.

Habitat Prefers still fresh water with fringing vegetation, inland and coastal. Resident in much of Europe including Britain but numbers here increase in winter.

Food and Habits A dabbling duck, eating mostly plants and seeds. May nest some distance from water. Fast in flight; springs up from water.

Mallard

Anas platyrhynchos

Female

Male

SIZE AND DESCRIPTION 58cm. Britain's most common duck. Drake has a dark rich brown breast, and a dark green head with a white collar in breeding season. Speculum (a bright, often iridescent patch of colour on the wings of some birds, especially ducks) is purple.

VOICE Ducks give the familiar 'quack'; drake has a higher-pitched call.

HABITAT Resident and widespread throughout Europe, occurring on almost any inland waters other than fast-flowing rivers. Often more coastal in winter.

FOOD AND HABITS Surface feeding, it can be seen dabbling and up-ending. Eats a variety of food, including invertebrates, fish and plants. Usually nests on the ground under bushes, close to water.

Pintail
Anas acuta

Size and description Female 56cm; male 66cm. Elegant male has a black tail, chocolate head and pure white breast. Female has a slim neck, dark bill, short pointed tail and overall brownish plumage.
Voice Rarely vocal; drake has a low whistle, female a low quack and churring growl.
Habitat Breeds in marshes and uplands in north-east Europe, very few in Britain. Winters on sheltered coastal and sometimes inland waters in Britain and further south and west.
Food and habits Dabbles and upends in shallow water. Winter diet includes aquatic plants, and roots, grains and other seeds. During nesting season eats mainly invertebrate animals. Gregarious, forming flocks with other ducks outside breeding season.

Garganey

Anas querquedula

Size and description 40cm. Slightly larger but slimmer than Teal, with longer bill. Male has broad white supercilium on dark brown head, breast and back brown, flanks silver, elongated tertials. Female and juvenile mottled brown with white belly and boldly striped face. All show plain greyish forewing in flight.

Voice Usually quiet, male has a dry rattle, female a soft quack.

Habitat Commoner in south. Prefers undisturbed, well-vegetated marshy ponds, flood-meadows and ditches. Uses more open water on migration.

Food and habits Summer visitor to Europe, winters in Africa. Dabbles and upends for aquatic invertebrates and vegetation. Discreet when breeding. Usually seen singly or in pairs in Britain.

Shoveler

Anas clypeata

Size and description 51cm. Surface-feeding duck easily recognized by its very large spatulate bill. Drake has a dark green head, white breast and chestnut flanks. Forewing is blue. Speculum is green.

Voice Drake calls 'took-took'; females 'quack'.

Habitat Scarce breeding bird in Britain, more numerous in winter as migrants from north-east Europe arrive. Well-vegetated freshwater wetlands, both inland and coastal.

Food and habits Feeds in shallow muddy water; sieves seeds through bill. Nest a down-lined grass cup well hidden on the ground.

Pochard

Aythya ferina

Female

Male

Female

Male

Size and description 46cm. Drake has a chestnut head, black breast, and grey back and flanks. Light-blue bill. Female is brown with a pale throat.

Voice A quiet bird; male gives a soft whistle, female growls.

Habitat Scarce breeding bird in Britain, more numerous in winter as visitors from further north and east Europe arrive. Breeds on well-vegetated freshwater wetlands, inland and coastal, in winter visits more varied waters including park lakes and reservoirs.

Food and habits A diving duck more active at night than during the day, and often seen resting on the water by day. Nest a down-lined grass cup well hidden on the ground.

Tufted Duck

Aythya fuligula

Size and description 43cm. A jaunty little diving duck. Drake is black and white with a drooping crest on the back of the head; duck is dark brown with the suggestion of a crest. Bill is blue with a dark tip.

Voice Tends to be silent.

Habitat Medium-sized or large fresh waters with fringing vegetation. More widespread in winter, when it occurs on more open gravel pits and reservoirs without cover. Widespread in Europe, wintering south to Mediterranean.

Food and habits Dives deeper than Pochard, eating mostly insects and molluscs. Nests on the ground a few metres from the water's edge.

Scaup
Aythya marila

Size and description 46cm. Large diving duck with a smoothly rounded head. Male has dark green-glossed head, black breast, wingtips and tail, and silver back and flanks (slightly darker above). Female grey-brown with mottled grey flanks and large white patch around bill base. Both sexes have yellow eyes and a small black 'nail' tip to grey bill.
Voice Quiet, soft purrs and growls in courtship.
Habitat Arctic breeder, winter visitor to north-west Europe. Usually seen offshore, in sheltered bays and estuaries, occasionally on lakes and gravel pits, commoner further north.
Food and habits Makes lengthy dives for food, particularly mussels. Will flock with other diving ducks.

Common Eider

Somateria mollissima

Size and description 60cm. The male of this large sea duck is unmistakable in breeding plumage. Heads of both sexes have a characteristic wedge shape with a long triangular bill. Non-breeding male may appear almost black, or chequered black and white. Female has brown-barred plumage.

Voice Male makes a surprised-sounding 'ah-hoo', female stuttering 'kokokok' calls.

Habitat Resident. Sea and rocky coastal areas. Much more common in north and Scotland and Ireland, where it breeds, than in southern Britain. Very rare inland.

Food and habits Dives for crustaceans and molluscs. Nests on coastal islands in colonies of 100–15,000 individuals.

King Eider
Somateria spectabilis

Size and description 55cm. A little smaller than Common Eider. Male has pale blue crown and greenish cheeks separated by thin white line, pink bill and very large yellow knob at bill base, white breast and black body with white flank patch and small black 'sail' on wing. Female resembles a smaller-billed female Common Eider.
Voice Cooing and croaking calls in courtship.
Habitat Breeds in high Arctic, very rare winter visitor to Britain. At sea, most likely in far north/north-east, invariably discovered among flocks of Common Eiders.
Food and habits Feeds on mussels, obtained by diving. Same individuals often return year after year.

Long-tailed Duck
Clangula hyemalis

Size and description Male 58cm, female 45cm. Compact seaduck, male with very elongated central tail feathers. Male in summer has black head, breast, wings and tail, white eye patch and belly brown upperparts, in winter much whiter with bold black cheek patch. Female also varies through year but always shows pale face with dark crown and cheek patch.

Voice Male has loud yodelling call, female a quiet quack.

Habitat Winter visitor, breeds in Arctic, a few oversummer in north Scotland. At sea, occasionally on coastal lakes or gravel pits.

Food and habits Dives with a small jump, eats molluscs and crustaceans. Gregarious and very active.

Common Scoter

Melanitta nigra

Size and description 48cm. A sea duck. Male is black with a patch of bright yellow on the beak. Female is dark brown with pale cheeks.
Voice A whistling 'pheeuu' in flight and while displaying.
Habitat Most likely to be seen on the sea, and sometimes on large inland reservoirs. Breeds in north-east Europe, and winters in North Sea, Baltic and Atlantic. In Britain occasionally breeds on remote lochs in Scotland and Ireland; otherwise found mainly off coasts in winter.
Food and habits Diet consists mostly of molluscs, and insects and fish eggs in freshwater habitats. Dives to 30m to hunt for shellfish. Congregates in large bobbing rafts of hundreds or thousands of birds well offshore.

Surf Scoter

Melanitta perspicillata

Size and description 46cm. Male solid black with white patches on forehead and nape. Large bill with swollen-looking base, yellow near tip and white at base with black spot. Female also large-billed, dusky grey-brown with large whitish patches in front of and behind the eye.
Voice Unlikely to call when in Britain, may make whistles and croaks in courtship.
Habitat A very rare winter visitor to Britain from North America. At sea, preferring sheltered bays and estuaries, most frequent in far north and north-east.
Food and habits Like other scoters, feeds mainly on mussels obtained by deep dives. Almost invariably found within rafts of other scoter species.

Velvet Scoter

Melanitta fusca

Size and description 55cm. Larger than Common Scoter. Male has black plumage with small white eye-flash, female dark grey-brown with whitish patches behind eye and at bill base. At a distance, both sexes most easily told from Common Scoter by white band on secondaries, revealed in flight or when they flap.

Voice Quiet when not breeding.

Habitat Breeds in Scandinavia, winter visitor to Britain, most numerous along east coast but scarcer than Common Scoter. Very rare inland.

Food and habits Dives to pull mussels from the sea bed. Gregarious, often flocks with Common Scoters, 'rafts' moving with the tides to stay near food sources.

Goldeneye

Bucephala clangula

Size and description 46cm. A diving duck. Conspicuous male is bright white and black with a glossy dark green head that has a circular white patch below the eye. Female and juvenile are grey with a brown head. In flight, makes a whistling noise with the wings.

Voice Rarely vocal. Male sometimes makes a disyllabic nasal call, female a harsh growl.

Habitat Found almost equally on coastal and inland waters. Mainly a winter visitor to Britain, also breeding occasionally in Scotland.

Food and habits Diet consists mainly of aquatic invertebrates, as well as amphibians, small fish and some plant material (mainly in autumn). Nests in hollows of mature trees.

Smew
Mergellus albellus

Size and description 42cm. Small, compact sawbill duck. Male white with black back, eye-patch and narrow breast-stripes, silver-grey flanks. Female and immature grey with rich chestnut crown and nape, cheeks white. Bill shorter and chunkier than those of other sawbills.
Voice Rarely calls. May give grunt or rattle.
Habitat Breeds in north-east Europe and Russia, winter visitor to western Europe. Freshwater coastal lakes and gravel pits, sometimes further inland. Most frequent on south and east coasts.
Food and habits May form small flocks. Makes long dives, takes fish and some insect larvae. Most seen in Britain are female-like (adult females and first-winters of both sexes).

Red-breasted Merganser

Mergus serrator

Size and description 55cm. Striking sawbill similar to Goosander, but less common. Male black above with flanks finely marked with grey, a white belly and a chestnut breast specked with dark brown. Head and crest are dark green, the throat white and the nape black. Female and juvenile are grey above, with a white belly and chestnut breast, nape and spiky crest.

Voice Generally silent.

Habitat Breeds beside rivers and lakes, and along sheltered coasts in northern Europe. Winters south to Black Sea. In Britain breeds in north and found around coasts in winter.

Food and habits Feeds mainly on small fish; also small amounts of plant material and aquatic invertebrates. Nest down-lined in hollow or burrow.

Goosander

Mergus merganser

Size and description 62cm. Largest sawbill. Male is vivid white with a dark green head that looks black from a distance, and a slender hooked bright red bill. Female and juvenile are grey with a white breast and a brown head that has a slightly shaggy crest.

Voice A quacking 'orr' during display.

Habitat Lakes, rivers and shores. In Britain resident in north and winter visitor in south.

Food and habits Diet consists mainly of fish, which it may pursue by swimming short distances. Nest down-lined in burrow or hollow tree.

Red-throated Diver

Gavia stellata

Size and description 60cm. Distinctive breeding plumage includes a brick-red throat, and a grey head and neck. At other times mainly grey-brown above, with white on the face that extends to above the eye. Slender upturned bill; swims with head mostly tilted upwards.

Voice Song an eerie wailing, heard in breeding areas. Flight call a goose-like 'ak ak'.

Habitat Breeds on lakes in northern Europe, including Scotland, and winters on coasts and large inland lakes.

Food and habits Primarily a fish-eater; may also feed on molluscs, crustaceans, frogs and plant material. Spears prey underwater, diving to 2–9m in depth. Nest is always close to water.

Black-throated Diver

Gavia arctica

Size and description 70cm. Intermediate between Red-throated and Great Northern Divers in size, bill size and bulk. In summer has smooth grey head, black throat, striped necksides and chequered back. In winter dull blackish-grey above, white below. Holds head level.
Voice Breeding pairs give loud wailing song. Silent in winter.
Habitat Breeds in northern Europe (in Britain very rare, north-west Scotland only), moves south in winter. Nests on islands in undisturbed lochs. In winter at sea and on deep coastal lakes and gravel pits. Occurs along all coasts in winter but is scarce.
Food and habits Feeds on fish, makes long deep dives. Usually solitary or small flocks.

Great Northern Diver

Gavia immer

Size and description 80cm. Largest diver, with bulbous forehead and heavy, dagger-like bill. In breeding plumage has black head with striped white neck patches, body dark with white chequering on back, bill black. In winter dark grey above, whitish below, bill mid-grey.
Voice Unlikely to be heard in Britain. Courting pairs have a wailing song.
Habitat High Arctic breeder, moving south to north-west European coasts in winter. At sea in sheltered bays, harbours and estuaries, sometimes reservoirs and lakes. Occurs on all coasts, some oversummer in far north-west.
Food and habits Dives for fish, swims buoyantly with level head, often rolls to preen. May form small groups.

Little Grebe
Tachybaptus ruficollis

Size and description 27cm. The Little Grebe, or Dabchick, in breeding plumage has bright chestnut cheeks and throat, and dark brown upperparts. In winter it is grey, but still has the abrupt 'powder-puff' rear. Sexes are similar.

Voice Whinnying song.

Habitat Resident in Britain and much of Europe. Mainly on inland pools, lakes and wide ditches, visits larger and more coastal waters in winter.

Food and habits Dives for food, mostly small fish. Rather skulking. Nests among waterside vegetation such as rushes, or under overhanging branches.

Great Crested Grebe

Podiceps cristatus

Size and description 48cm. Unmistakable in breeding plumage; both sexes have a large horned crest and ruff, which are lost in winter. Chicks are striped.

Voice Generally silent. Call usually a harsh bark; crooning song.

Habitat Breeds on still waters, lakes, ponds and reservoirs, and slow-flowing rivers. May be found on coasts and estuaries in winter. Occurs in much of Europe except far north.

Food and habits Dives for food, which consists mostly of fish and invertebrates, and some plant matter. Often ingests feathers. Nests on a floating mat of reeds and other water plants near the water's edge.

Red-necked Grebe

Podiceps grisegena

Size and description 45cm. Slightly smaller and stockier than Great Crested Grebe, without obvious crest. In breeding plumage has black crown that reaches eye, white cheeks and red neck. In winter duskier with pale neck. Always shows yellow bill-base.

Voice Usually silent but has loud hooting, wailing song in courtship.

Habitat Breeds on well-vegetated lakes, sometimes among gull colonies. Mainly breeds in eastern Europe and beyond, very rare and sporadic breeder in Britain, most likely on east coast. In winter, in sheltered seas, also lakes. Scarcest grebe species.

Food and habits Dives for fish and other aquatic prey. Swims buoyantly and dives energetically. Usually solitary or in small groups.

Slavonian Grebe

Podiceps auritus

Size and description 35cm. Larger and longer-necked than Little Grebe. Rather flat head profile with small peak near rear of the crown, bill shortish and straight, eye red. In summer colourful, with black head and back, bright rufous neck and underside, and broad yellow ear-tufts. In winter blackish above, white below.

Voice Trilling call when breeding, quiet in winter.

Habitat Breeds in north-eastern Europe and beyond, including a few in north Scotland, on small, lushly vegetated, shallow lochs and lochans. In winter occurs on sheltered seas and coastal lakes, rarely inland.

Food and habits Dives for aquatic insects and small fish. Loosely colonial when breeding, and may form small parties on migration and in winter.

Black-necked Grebe

Podiceps nigricollis

Size and description 32cm. Small, slim-necked grebe, often puffy rear end. Head shape steep at forehead with peaked central crown, short bill looks uptilted. Prominent red eye in all plumages. In summer has black head, back and breast, frill of yellow ear-tufts. In winter black above, white below with dusky grey neck and flanks.
Voice Trills and whistles when breeding, otherwise silent.
Habitat Breeds patchily across Europe, including sporadically in Britain. Nests on islands on shallow lakes, sometimes colonially. In winter on coastal and inland lakes, sometimes sheltered seas.
Food and habits Dives for aquatic invertebrates and small fish. Social both in summer and winter.

Fulmar

Fulmarus glacialis

Size and description 47cm. Seabird that resembles a gull, usually having grey upperparts and white underparts, but with a rather thick neck. Wings are held straight and stiff in flight. Nostrils are located in short tubes halfway down the bill.

Voice Guttural chuckles and growls mainly at nest.

Habitat Breeds around whole British coast, more pelagic in autumn and early winter. Coasts, cliffs and sea.

Food and habits Eats fish, offal and molluscs. Nests in colonies on cliff faces. Returns to breeding sites from November onwards, dispersing in late summer. Flight interspersed with frequent glides.

Cory's Shearwater

Calonectris diomedea

Size and description 45cm. Large, rather pale shearwater, with very long wings, strikingly pale yellowish bill, upperside drab grey-brown, underside white, underwing shows contrasting dark flight feathers. Has relaxed flight action, alternating gull-like flapping with long banks and shears.

Voice Silent at sea.

Habitat Breeds in Mediterranean, wanders widely over open sea in winter, especially Bay of Biscay. Very rare in Britain, most likely to be seen from south-western headlands in autumn.

Food and habits Mainly picks food from on or near the surface but also makes shallow plunge-dives, taking carrion and small fish and squid. Will follow boats. Moves closer to land during rough weather at sea.

Manx Shearwater *Puffinus puffinus*

Size and description 32cm. Small, delicate-looking, long-winged seabird, plumage dark grey-black above, white below, underside of wings white with narrow dark border. Can resemble an auk in flapping flight when wing length less obvious.
Voice Silent at sea. Bizarre howling calls at breeding grounds.
Habitat Breeds in burrows on islands, most of breeding population is in northern and western Britain. After breeding wanders widely at sea, may be seen from headlands anywhere around Britain.
Food and habits Feeds on small marine animals and floating carrion, mainly picked from surface. Flight alternates fast flapping with shearing. Visits breeding sites only at night. Gregarious at all times of the year.

Sooty Shearwater *Puffinus griseus*

Size and description 45cm. Very dark, long-winged, medium-sized seabird. Wings on underside show pale silvery inner lining, plumage otherwise uniform sooty grey-black. Tail wedge-shaped, bill slender. Could be confused with a dark skua, but smaller with proportionately longer, narrower wings and distinctive flight style.
Voice Silent at sea.
Habitat Breeds in the southern hemisphere on islands, roams very widely after breeding with some passing north Atlantic coasts. Rather scarce in Britain, most often seen from south-western headlands in late summer.
Food and habits Feeds on small marine animals and floating carrion, mainly picked from surface. Powerful shearing and banking flight action.

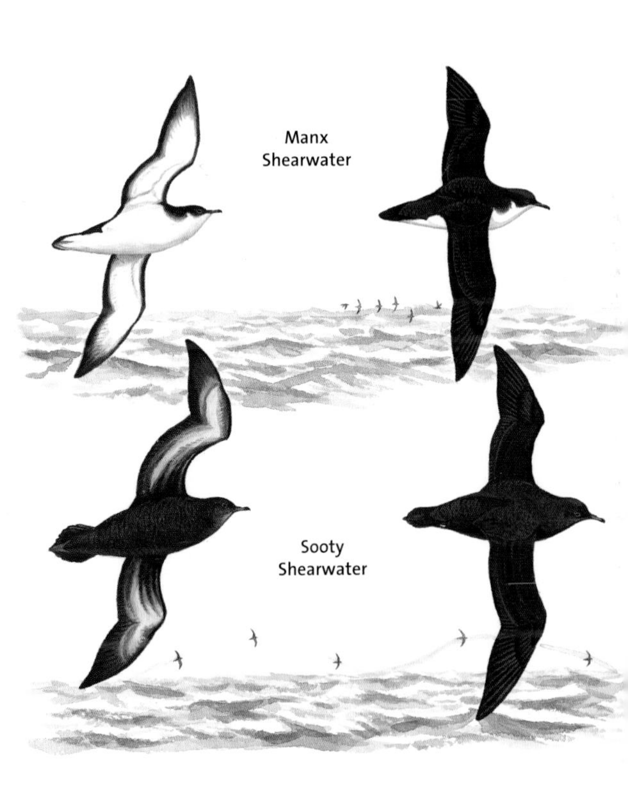
Manx
Shearwater
Sooty
Shearwater

Great Shearwater

Puffinus gravis

Size and description 46cm. Large shearwater, resembling oversized and browner version of Manx Shearwater. Upperside dark brown with paler feather fringes giving scaly look, narrow pale collar and rump patch. Underside white, underside of wings white with dark outline. Flight powerful and direct, shearing on stiff wings close to wave crests.

Voice Silent at sea.

Habitat Breeds in south Atlantic, ranges very widely at sea in northern summer. Very rare, most often seen from south-western headlands in autumn.

Food and habits Similar to Cory's Shearwater. Both the large shearwaters are gregarious by nature but few wander as far north as Britain so tend to be seen singly.

Balearic Shearwater

Puffinus mauretanicus

Darker variant

Size and description 35cm. Slightly larger and much duskier than Manx Shearwater, upperside dark grey-brown and underside only slightly paler, with no crisp dividing line. Shows white in centre of underwing 'hand'. Flight a little stronger and less fluttering than Manx.

Voice Silent at sea.

Habitat Breeds on Balearic islands but ranges widely at sea in winter. Most often seen from southern and eastern coasts.

Food and habits Takes fish, squid and carrion. Typical shearwater feeding behaviour – surface-picks, makes shallow dives, and follows fishing boats. Declining severely and classed as Critically Endangered, its breeding colonies threatened by holiday resort development, and predation by cats and rats.

European Storm-petrel *Hydrobates pelagicus*

Size and description 15cm. Tiny blackish seabird with square white rump patch and whitish wingbar. Long, slender legs, square-cut tail, long, round-ended wings, steep-fronted head and small bill with prominent tubular nostrils.
Voice Silent at sea, purring and grunting calls at nest.
Habitat Breeds in burrows or rock crevices on islands in a few parts of northern and western Britain. After breeding travels long distances at sea, and may be seen from headlands anywhere around Britain.
Food and habits Picks tiny marine animals and carrion scraps from surface. When feeding often 'patters' feet on sea surface. Will follow fishing boats. Visits breeding sites only at night.

Leach's Storm-petrel *Oceanodroma leucorhoa*

Size and description 21cm. Larger than Storm Petrel with longer, more pointed wings giving more dynamic flight, slightly paler plumage. Rump patch smaller and narrower, tail has slight fork. Wingbar broad, smoky grey.
Voice Silent at sea, purring and grunting calls at nest.
Habitat Breeds in burrows or rock crevices on islands in extreme north-west Britain and elsewhere in Europe, also North America. After breeding wanders widely at sea, most often seen from western headlands in autumn, occasionally 'wrecked' inland after gales.
Food and habits Picks small marine animals and floating carrion from surface. Relatively strong flapping and shearing flight. Visits breeding sites only at night. Follows boats.

European
Storm-petrel
Leach's
Storm-petrel

Gannet
Morus bassanas

Size and description 90cm. Large seabird with long narrow wings that have black tips, a pointed tail, a long and pointed blue-white bill, and a yellow tinge to the back of the neck. Juveniles are dark, becoming lighter as they mature at three years of age.
Voice Harsh croaks at nest.
Habitat Entirely maritime, only coming to land to breed. In Britain breeds on northern and western coasts in Scotland and Wales. Can be seen offshore almost anywhere, especially when migrating south in autumn.
Food and habits Diet almost exclusively fish, which are caught by spectacular arrow-shaped plunge-dives from heights of 15–30m. Nests in dense colonies on cliffs and rocky islands.

Cormorant

Phalacrocorax carbo

Breeding

Adult non-breeding

Juvenile

Breeding

Size and description 90cm. Very dark seabird with a white throat and cheek patches, black-bronze upperparts and blue-black underparts. White thigh patch in breeding season, when some birds also have a white head. Swims low in the water. On land 'heraldic' pose with wings held out is characteristic. Sexes are similar; juvenile is brown.

Voice At nest makes guttural noises.

Habitat Present throughout the year on coast; sometimes inland on islands on lakes and rivers.

Food and habits Eats fish almost exclusively, catching them by diving. Nests in colonies, usually on rocks on coast.

Shag

Phalacrocorax aristotelis

Size and description 75cm. Similar to Cormorant except in full breeding plumage, when Shag has a greener sheen and a quiff on its head. There is also a thick yellow gape reaching beyond the eye, there are no white patches and the forehead is steeper than that of a Cormorant.

Voice Harsh croaks on breeding ground.

Habitat Year-round resident of rocky coasts and nearby seas. Very unusual inland, unlike Cormorant. Local along coasts of Europe, north-west Russia and North Africa.

Food and habits Feeds on fish, which are taken mostly by diving from the surface. Semi-colonial or solitary breeding bird.

Bittern
Botaurus stellaris

Adults

Size and description 75cm. Plumage brown marbled and striped with buff and black, offering good camouflage against dead reeds in its habitat. Freezes in an upright position when alarmed.
Voice In spring male utters a far-carrying booming 'woomb' or 'oo-hoo-oomb', like a foghorn, mainly at night. Call in flight a barking 'cow'.
Habitat Large freshwater reed beds year-round in much of central Europe. Rare breeding bird in Britain, mainly extensive coastal reedbeds in east England, more widespread in winter.
Food and habits Diet consists of fish, frogs, insects, small mammals and birds, and snakes. Hunts by walking slowly among plants, lifting its feet high with each step. Nest a reed platform among reeds. Declined alarmingly in the 1990s, but recent successful conservation measures have resulted in an encouraging increase in the population.

Little Egret

Egretta garzetta

Size and description 60cm. A very graceful white heron. Black bill and black legs with yellow toes. Long white plumes on nape and back in breeding plumage.

Voice Flight call is a harsh 'ktchar'.

Habitat Estuaries, marshes, rivers, saline lagoons and other shallow water bodies. Local in southern and central Europe. Increasingly common in Britain and Ireland since the 1990s, and now a common resident along coast in much of England and Wales.

Food and habits Diet consists of various animals such as small fish, amphibians and insects. Nests colonially in bushes near wetlands.

Great White Egret
Ardea alba

Size and description 90cm. Large, elegant white heron, similar size to Grey Heron but slimmer, with very long, slim, serpentine neck and proportionately short bill (yellow in winter, darkens in summer). Feet and lower legs black, legs paler close to body. Breeding birds have elongated fine plumes on lower back.

Voice Low-pitched grunts and caws in the breeding season.

Habitat Scattered breeding distribution across southern Europe. Scarce but increasing winter visitor to mainly southern Britain, occasionally breeds. Low-lying marshy wetlands. Nests in trees, bushes or reedbeds.

Food and habits Takes fish, frogs and other small animals. Stalks prey or waits motionless before striking. May roost with other herons.

Grey Heron
Ardea cinerea

Size and description 95cm. Very large and mainly grey, with black-and-white markings. Breeding plumage includes long black plumes on head. Neck is tucked back in flight; wingbeats are slow and ponderous.

Voice Flight call a hoarse croaking 'kraark' and 'chraa'; bill-clapping at nest.

Habitat Year round in marshes, ponds, lakes, rivers, canals, flooded fields and estuaries throughout Europe.

Food and habits Feeds on fish, amphibians, small mammals, insects and reptiles. Hunts by stalking slowly through shallow water, or standing motionless waiting for prey to come within reach, when it strikes with lightning speed. Nests in colonies, usually high in tall trees, in a huge nest.

Spoonbill
Platalea leucorodia

Size and description 85cm. Unmistakable large white bird with a long flat bill that broadens at the tip.

Voice Generally silent.

Habitat On the Continent breeds in large reed beds around shallow wetlands in south and the Netherlands. Winters in western Europe and Africa. In Britain most likely at one of east coast nature reserves in spring or autumn; in winter a few mainly on southern estuaries.

Food and habits Feeds on molluscs by sieving water with side-to-side head movement. Nests colonially in platform nests erected in large reed beds. Rare in Britain and of European conservation concern.

White-tailed Eagle

Haliaetus albicilla

Size and description 85cm. Largest British raptor. Rectangular wing shape giving 'barn door' flight outline. Head large, tail short. Adult brown, paler on head, tail white, huge bill yellow. Juvenile all-dark. Flies with long glides and heavy, heron-like flapping.
Voice Sharp yapping call, mainly near nest.
Habitat Breeds patchily across central Europe. In Britain only in north Scotland and Ireland. Rocky coastlines, large lochs, low-lying farmland. Nests in large trees or on crags, same sites in constant use for decades.
Food and habits Feeds mainly on fish, caught in low swoop and talon swipe, also takes seabirds and much carrion. Often in pairs. Many other birds mob them relentlessly.

Marsh Harrier

Circus aeruginosus

Size and description 52cm. Largest European harrier. Usually dark brown above with buff shoulders and head. Male has a grey tail and grey secondaries. Often flies with wings held in a shallow 'V'.
Voice Two-note display call, 'kweeoo'.
Habitat Reedbeds and coastal marshes. Scarce but increasing breeding bird in Britain. Resident in more southerly parts of Europe, including southern England, summer visitor further north.
Food and habits Eats small mammals, frogs, and birds and their eggs. Hunts by systematically quartering the ground. Nests in reed beds.

Hen Harrier
Circus cyaneus

Size and description 45cm. Slimmer build and narrower wings than Marsh Harrier. Male grey with white underparts.
Voice Display call 'tchik-ikikikik'.
Habitat Rare British breeding bird on moorland, mainly north Scotland and Ireland. Widespread on coastal farmland and marshes in winter.
Food and habits Diet includes small mammals, birds, reptiles, insects and carrion. Nest a grassy platform on the ground.
▶ **Similar species** **Montagu's Harrier** (*C. pygargus*). Slighter and rarer than Hen Harrier, with more pointed wings. Underparts speckled with brown. Voice higher than Hen Harrier's. Summer visitor breeding in a few sites mainly in eastern England.

Common Buzzard

Buteo buteo

Size and description 52cm. Large with broad rounded wings and a short tail. Usually dark brown above with variable amounts of white below; sometimes with a dark carpal patch.

Voice Mewing cry, 'peeioo'.

Habitat Moorland and agricultural land. Year-round resident across much of Europe; summer visitor to far north.

Food and habits Feeds mainly on small mammals, which it catches with a low-flying pounce; also carrion. Soars on V-shaped wings. Nest is a bulky structure of twigs, usually erected in a tree.

Similar species **Honey Buzzard** (*Pernis apivorus*), 54cm long, has a smaller head than the Common Buzzard, and soars on flat wings. Feeds mainly on wasps, digging out their nests. Summer visitor to much of Europe; scarce in Britain. Winters in Africa. See also **Rough-legged Buzzard** (page 70).

Rough-legged Buzzard

Buteo lagopus

Size and description 54cm. Slightly larger than Common Buzzard with longer, slimmer wings. Legs (but not feet) lightly feathered. Usually paler than Common Buzzard, with broad dark belly and carpal patches, tail pale with thick dark terminal band.

Voice High, wailing 'peeooooo'.

Habitat Actic breeder, uncommon winter visitor to Britain, mainly along east coast, numbers fluctuate from year to year. Mainly found on open, low-lying coasts with rough grassland, farmland or grassy marsh.

Food and habits Hunts small mammals and rabbits, also readily takes carrion. Searches for prey from soaring flight, often hovers. Sometimes waits and pounces from a low perch.

Osprey

Pandion haliaetus

Size and description 55cm. Large and graceful fish-eating bird of prey that has dark brown upperparts and white underparts. Long narrow wings are held angled in flight.

Voice Call a short shrill whistle.

Habitat Lakes, lochs and rivers; reservoirs on passage. Uncommon summer visitor to Britain; breeds in Scotland, and in smaller numbers in Wales and northern England. May be seen at almost any large body of fresh water during spring and autumn migration.

Food and habits Feeds on fish, which it catches by plunge-diving from 10–30m. Flies holding fish in talons in line with its own head and tail. Nests in tall trees.

Kestrel

Falco tinnunculus

Size and description 34cm. Distinctive long tail and pointed wings. Male has a grey head, black-tipped grey tail and dark-flecked russet back. Female and juvenile lack the grey head, and have a brown tail with narrow bars, and more dark flecks on the back.

Voice Noisy at nest-site; rasping 'kee-kee-kee-kee' call.

Habitat Farmland, moorland and other open areas. Breeds in cities and towns; may be seen flying over gardens. Resident across Europe; northern and eastern European populations migrate during autumn.

Food and habits Hovers above grassland or perches on trees and pylons, ready to drop down on rodents in grass. Also feeds on small birds, large insects and lizards. Lays eggs in a hole or on a bare ledge.

Merlin

Falco columbarius

Size and description 29cm. Small and compact falcon with short pointed wings. Male blue-grey above and buff with dark spots below, with indistinct moustaches. Female and juvenile brown above.

Voice Calls are 'kee-kee-kee'.

Habitat British breeding population is at the south-west limit of the species' European range; thinly scattered across upland moorland from south-west England north to Shetland. Often found on coast in winter. Nowhere common.

Food and habits Feeds on small birds, which it hunts in fast flight close to the ground. Nest usually on the ground, among heather.

Hobby
Falco subbuteo

Size and description 32cm. Dashing little falcon that looks like a large swift in flight. Dark slaty-grey above with dark moustaches on white cheeks and throat, and red thighs.

Voice A repeated clear 'kew-kew-kew'.

Habitat Summer visitor, nests mostly on heathland, commoner in south. Visits coastal wetlands before and after breeding, sometimes in groups of 10 or more.

Food and habits Feeds on small birds, and large insects such as dragonflies, which are often eaten in flight. Usually nests in an abandoned crow's nest.

Peregrine Falcon

Falco peregrinus

Adult

Juvenile

Size and description 45cm. Large and compact falcon with a heavy build. Adults have strong black moustaches and horizontal barring on underparts, and are bluish steely-grey above. Female larger than male. Largest British falcon.

Voice Calls are 'kee-kee-kee'.

Habitat Resident, nests on sheer cliffs and buildings both on coast and inland, often hunts gatherings of birds on estuaries and coastal marsh in winter.

Food and habits Feeds on birds, including feral pigeons. Circles high up waiting for prey to fly below, then plunges at high speed in pursuit. Eggs laid in a bare scrape.

Water Rail

Rallus aquaticus

SIZE AND DESCRIPTION 24cm. A secretive bird that is often hidden in reeds and more often heard than seen. Grey underparts, white-barred flanks, a red bill and a pointed tail that is usually held erect.

VOICE Pig-like squeaking and grunting, and a high-pitched 'kip-kip'. Male display call 'kurp kurp kurp', female 'tchik-tchik'.

HABITAT Resident in Britain and much of Europe, summer visitor in northern Europe. Marshes, reedbeds and other densely vegetated wetland habitats.

FOOD AND HABITS Omnivorous. Diet consists mainly of small animals such as worms, molluscs, shrimps, crayfish, spiders, insects, amphibians and fish; also feeds on plant matter. Nest a cup of vegetation, usually on a thick stand of reeds or rushes.

Spotted Crake

Porzana porzana

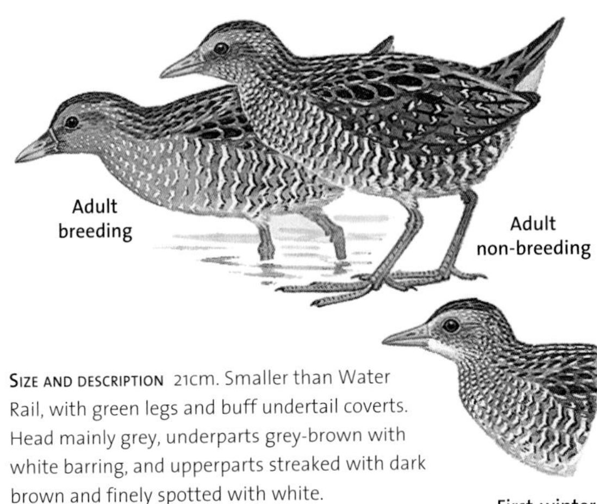

Size and description 21cm. Smaller than Water Rail, with green legs and buff undertail coverts. Head mainly grey, underparts grey-brown with white barring, and upperparts streaked with dark brown and finely spotted with white.

Voice Whip-like 'quip, quip, quip' call at night during breeding season.

Habitat Summer visitor to Britain and Europe, very rare breeding bird with scattered distribution. Marshes, wet lush meadowland and similar damp, well-vegetated habitats.

Food and habits Probes mud and shallow water with bill to pick up invertebrates; also hunts by sight. Nests in marsh vegetation.

Moorhen

Gallinula chloropus

Size and description 30cm. Distinctive slaty plumage, dark brown wings, white undertail coverts, yellow-tipped red bill and green legs. Flicks tail as it walks with a careful tread. Juvenile is brown.

Voice Varied repertoire includes harsh metallic 'krrek' and 'kittick' calls.

Habitat Resident and very common in Britain. Ponds, rivers, canals, lakes and marshes across Europe. Also parks and gardens with large ponds.

Food and habits Feeds on seeds, insects, molluscs, leaves and carrion. Nest a bulky mound of vegetation on the water. Juveniles may help parents raise next generation.

Coot
Fulica atra

Size and description 38cm. Mainly black water bird with a white bill and shield on the forehead, greenish legs and a domed back. Chicks are black with rufous heads. Juvenile is greyish.
Voice Quite noisy; call usually a loud 'kowk' or variation.
Habitat Resident and very common in Britain. Still and slow-moving fresh water. Usually found on larger and more open water bodies than Moorhen.
Food and habits Dives for food, largely aquatic plants. Often in flocks, especially outside the breeding season. Requires fringing vegetation for nesting. Quarrelsome; fights on the water using its large feet, especially during the breeding season, when it will attack birds much larger than itself such as swans and geese.

Oystercatcher

Haematopus ostralegus

Size and description 43cm. A large and boldly marked black-and-white wader with an orange-red bill and pink legs. White rump and wing bar, and white collar in winter.

Voice Noisy; loud 'kleep' call and piping display.

Habitat Coasts, mudflats and wet meadows; usually coastal. Resident in Britain; also passage migrant and winter visitor.

Food and habits Eats mainly invertebrates, especially molluscs, which it opens by hammering or prising (some Oystercatchers have pointed bills, others have squarer-ended bills). Nests on the ground.

Black-winged Stilt
Himantopus himantopus

Size and description 35cm. Unique, unmistakeable wader, with slim tapered body balanced on preposterously long, slim pink legs. Head and body white, wings black, develops variable dusky crown and back of neck when breeding. Bill black, straight and shortish.
Voice Noisy when breeding, especialy if nest is threatened, gives volley of shrill barking calls.
Habitat Shallow marshy wetlands in southern Europe. Rare visitor to mainly southern Britain, occasionally breeds.
Food and habits Picks tiny insects, crustaceans and other aquatic invertebrates from on or in shallow water. Gregarious, nests in colonies in its usual breeding range, very aggressive to other birds that approach its nest.

Avocet

Recurvirostra avosetta

Size and description 43cm. Large, elegant and boldly marked black-and-white wader with a black crown and nape, blue-grey legs and a slender upturned bill.

Voice Call a liquid 'kluut'.

Habitat Salt marshes, brackish lagoons and mudflats. Mainly summer visitor to Britain, breeding mostly in East Anglia and south-east England. Winters in south-west and west Africa, but many birds stay in western Europe in mild winters.

Food and habits Feeds chiefly on shrimp-like crustaceans, ragworms, other invertebrates and fish spawn, using vigorous sideways swishes of bill through water and silty mud. Nest lined with marsh vegetation in a shallow scrape on the ground.

Little Ringed Plover

Charadrius dubius

Size and description 16cm. Very similar to Ringed Plover, but less common, and adults have a golden eyering, duller straw-coloured legs and a black bill. No narrow white wingbar.

Voice Calls a whistling 'tiu'. Display or flight call a rough rolling 'chrechrechrechre'.

Habitat Inland marshes, lakes and gravel pits. Summer visitor to most of Europe. In Britain fairly common in south, but absent from north.

Food and habits Diet consists mainly of insects, as well as spiders, freshwater shrimps and other small crustaceans. Nest a shallow scrape on loose sand, dry mud or rocks, or in sparse vegetation, near water.

Ringed Plover
Charadrius hiaticula

Size and description 19cm. Common shore bird with a black mask and breast band, and a white collar and forehead. Upperparts brown, underparts white. Narrow white wingbar. Black markings more subdued in winter than in summer.

Voice Calls a liquid 'tooi' and 'kluup'. Trilling song.

Habitat Resident in Britain. Breeds on beaches and mudflats; also lake edges or tundra, sometimes inland, in north. Winters mainly on rocky and muddy coasts in western Europe.

Food and habits Mainly eats insects, worms and molluscs, and some plant matter. Nests in a shallow scrape on the ground.

Golden Plover
Pluvialis apricaria

Size and description 28cm. Breeding plumage spangled yellow and brown, with black throat and belly, northern birds being more boldly marked. Outside breeding season black on underparts is absent.
Voice Liquid whistling call, 'tlui'. Song a liquid 'too-roo, too-roo'.
Habitat Rare breeding bird of upland moor, mainly in northern Britain. In winter widespread and much commoner, mainly found on coastal grassland and grazing marsh.
Food and habits Eats mostly insects, molluscs and some plant matter. Nests in a grass-lined scrape well hidden on the ground.

Adult winter

Adult summer

Grey Plover
Pluvialis squatarola

Size and description 28cm. A strikingly handsome plover. Adult in summer has silver-grey black-flecked upperparts, separated from the black face, throat, breast and belly by a wide margin. Winter adult and juvenile paler and duller. Bill and legs black.
Voice Call a plaintive three-syllable 'plee-oo-ee'.
Habitat Mudflats and estuaries along coasts of North Sea, Atlantic and Mediterranean. Breeds on northern tundra. In Britain common on coast and present all year except midsummer.
Food and habits Feeds mostly on insects, and some plant matter in breeding season; on marine polychaete worms, molluscs and crustaceans in winter. Nest a shallow scrape on dry ground in an exposed stony site.

Lapwing

Vanellus vanellus

Size and description 30cm. Dark and glossy metallic-green upperparts, white below with a buff undertail, and a long wispy crest. Throat black in breeding season. Juvenile has a short crest. Floppy, loose, broad-winged flight. Tumbling display flight by males in spring.

Voice Calls 'peewit'.

Habitat Breeds on farmland and marshy grassland both inland and on coast in Britain and western Europe, widespread. In winter, distribution more coastal.

Food and habits Diet includes insects, worms and molluscs, with some vegetable matter. Nests on the ground. Winter flocks may be mixed with Golden Plovers.

Knot

Calidris canutus

Size and description 25cm. A little larger than Dunlin, with a comparatively short bill. Winter plumage is grey. Narrow white wingbar. Breeding plumage is rufous.

Voice Call 'knut'.

Habitat Breeds on High Arctic tundra. Widespread on mudflats on North Sea and Atlantic coasts during migration and winter.

Food and habits Feeds on invertebrates such as insects, molluscs, earthworms and crustaceans. Usually seen in flocks, which can be very large and dense. Spectacular aerial manoeuvres as flocks come in to roost. Nest a grass-lined scrape well hidden on the ground.

Sanderling

Calidris alba

Size and description 20cm. Pale grey in winter plumage. Summer plumage scaly brown above and on the breast; white belly shows white wingbar in all plumages.

Voice In flight, liquid 'twick, twick'.

Habitat Breeds further north than Britain, where it is a passage migrant and winter visitor confined to coast.

Food and habits Distinctive feeding method as it runs in and out on the shore with the movement of the waves, often likened to a clockwork toy. Nest a grassy cup well hidden on the ground.

Little Stint

Calidris minuta

Size and description 15 cm. Tiny wader with black legs and a short, fine black beak. Winter plumage (rarely seen in Britain) soft grey with white underparts. Breeding plumage rufous on head and breast, turning buff-brown by July.

Voice Call a feeble short 'pip'.

Habitat Breeds on tundra in far north. On passage widespread on marshes and mudflats in coastal areas of Europe.

Food and habits Feeds mainly on insects; also crustaceans and molluscs.

Temminck's Stint

Calidris temminckii

Size and description 14cm. Shorter-legged with more horizontal and hunched posture than Little Stint, plumage pattern similar to that of Common Sandpiper with grey-brown upperside, white below and white 'spur' in front of wing bend. Narrow white wingbars and tail-sides show in flight. Bill short and straight, legs dull yellowish, tail relatively long.

Voice Call a soft trill.

Habitat Arctic breeder, in Britain rare passage migrant, tiny numbers breed in north Scotland. Small pools, lakes, ditches and marshes, both coastal and inland.

Food and habits Feeds on insects found at the water's edge. Moves relatively slowly and deliberately, if disturbed takes off near-vertically and zigzags away in Snipe-like manner.

Pectoral Sandpiper
Calidris melanotos

Juvenile

Size and description 19cm (female), 22cm (male). Resembles larger, small-headed Dunlin. Mottled grey-brown on upperside, face and breast, with abrupt transition from streaked breast to clean white belly. Pale supercilium. Bill shortish, slightly downcurved, black with yellowish base, legs yellowish.

Voice A rasping 'kreet' given in flight.

Habitat Rare visitor from North America, more often on western coasts, the most frequent North American vagrant wader. Some make lengthy stays. Usually by freshwater, on lake shores, creeks, sometimes reservoirs.

Food and habits Mainly insects and other invertebrates. Has less frantic feeding action than Dunlin. Usually associates with parties of other waders.

Adult summer

Curlew Sandpiper

Calidris ferruginea

Size and description 20cm. Elegant sandpiper with a long curved bill, and a long neck and legs. Breeding plumage (rarely seen in Britain) includes a striking deep rufous colour on the breast and head. Juvenile has neatly scaled sandy brown upperparts and white flanks.

Voice Call a disyllabic 'chirrip'.

Habitat Breeds on Russian High Arctic tundra. Migrates via Europe to Africa. In Britain most likely to be seen on passage in spring or autumn.

Food and habits Diet includes snails, worms and insects. Forages by probing mud rapidly with bill.

Purple Sandpiper
Calidris maritima

Size and description 21cm. Compact and rotund smallish wader. Plumage dark dusky grey, paler on underside but with heavy grey streaking on breast and flanks. Legs short, bright yellow-orange. Bill fairly short, slightly downcurved, dark with orange base. Narrow pale wingbars in flight.

Voice Quiet but flocks in flight may give a musical twitter.

Habitat Arctic breeder, passage migrant and winter visitor to Britain, a handful breed in Scotland. Rocky shores, more rarely beaches, often roosts on breakwaters or other man-made structures.

Food and habits Searches rocks at low tide for molluscs, crabs and other invertebrates, flutters up to avoid breaking waves. Often associates with Turnstones. Can be confiding.

Dunlin
Calidris alpina

Size and description 19cm. Slightly smaller than Sanderling, with longer bill and less conspicuous white wingbar. Summer plumage scaly black and brown above, and white below with a large black belly patch. In winter it is greyer with a whitish belly.

Voice Call in flight 'treep'.

Habitat Breeds on northern and Arctic tundra, a few pairs breed in north Scotland. On migration and in winter is the commonest small wader on British coasts, found on flat beaches, estuaries and coastal marshland.

Food and habits Feeds on small invertebrates such as molluscs and worms. Nest a grassy cup well hidden on the ground.

Ruff

Philomachus pugnax

Size and description Male 30cm; female 23cm. Large male has a bright ruff (or partial ruff) of feathers in summer. Female, winter male and juvenile much duller. Legs long and orange; bill slightly decurved, and orange and black in male, blackish in female.

Voice Usually silent.

Habitat Breeds central Europe, very rare British breeder (mainly eastern England). Common passage migrant on coastal lagoons and marshes, a few overwinter.

Food and habits Diet is mainly insects, crustaceans, molluscs, amphibians, small fish, and cereals and aquatic plants. Nest a deep cup well hidden in grass.

Jack Snipe

Lymnocryptes minimus

Size and description 19cm. Markedly smaller and shorter-billed than Snipe, with more yellow tones to very streaky, well-camouflaged plumage. No pale central crown stripe, but has short, narrow, dark 'eyebrow' line through the broad yellowish supercilium.

Voice Quiet, unlike Snipe rarely calls when flushed.

Habitat Breeds in north-east Europe, Winter visitor to western Europe, widespread. Wet, muddy, marshy fields, ditches and lake shores.

Food and habits Probes soft ground for worms, also takes insects and other invertebrates. Has characteristic constant bobbing action, as if spring-mounted, whether walking or standing. Very shy and discreet, reluctant to flush, and escapes with direct rather than zigzag flight.

Adult

Common Snipe

Gallinago gallinago

Size and description 27cm. Wader most likely to be seen when flushed, flying off in a zigzag fashion. Extremely long bill, striped yellow, and dark brown head and upperparts.

Voice Hoarse cry when flushed.

Habitat Breeds on moorland and wet grassland in Europe, in Britain commoner in north. In winter common and widespread on coastal marsh and other damp, well-vegetated habitats.

Food and habits Eats mainly worms; also molluscs, insects and other invertebrates. Display flight involves a 45-degree dive, with a bleating noise caused by air rushing through outspread tail feathers. Nest a deep cup well hidden in grass.

Black-tailed Godwit

Limosa limosa

Size and description 41cm. Breeding bird has a rufous-coloured breast. Colour outside breeding season grey-brown. In flight, a broad white wingbar and a white band on the tail above a black band distinguish it from Bar-tailed Godwit.

Voice Call in flight 'wicka-wicka-wicka'. Song 'crweetuu'.

Habitat Breeds on grassland and flood meadows. Winters in sheltered coastal areas in southern and western Europe. In Britain a scarce breeder in East Anglia and a widespread winter visitor.

Food and habits Feeds mostly on insects and larvae; also molluscs and worms. Nest a cup well hidden in a tussock.

Bar-tailed Godwit

Limosa lapponica

Size and description 37cm. Shorter legged and more robust than Black-tailed Godwit. In flight shows a distinct white rump and a barred tail.
Voice Flight call nasal, similar to Knot's.
Habitat Breeds in Scandinavia and on tundra. Winters on tidal mudflats and sandy shores from Britain southwards.
Food and habits Probes mud for crabs, shrimps and marine worms in winter; insects taken mainly in summer. Nest a well-concealed scrape on the ground.

Whimbrel

Numenius phaeopus

Size and description 41cm. Smaller and more slender than Curlew, with a shorter bill and a markedly striped face, pale crown stripe and dark stripe above the eye. White 'V' shape on rump shows in flight.
Voice Liquid bubbling call.
Habitat Rare upland breeding bird in northern Scotland, fairly common passage migrant to rest of Britain, occurs on estuaries, marshes, and lake shores both coastal and inland.
Food and habits Eats mostly molluscs, worms and crustaceans. Nests in a well-hidden grassy cup on the ground.

Curlew

Numenius arquata

Size and description 54cm. Largest wader, with a very long decurved bill. Plumage streaked brown. Bigger and more robust than Whimbrel. White 'V' shape on rump shows in flight.

Voice Distinctive liquid call, 'coor-wee'.

Habitat Breeds on moors and wet meadows, and winters on mudflats and fields, often on coasts. Resident in Britain.

Food and habits Eats mainly small invertebrates, fish and plant matter. Long bill enables it to probe mud and sand deeply. Occurs in flocks outside breeding season, but feeds more separately. Nest a grassy cup well hidden on the ground.

Spotted Redshank
Tringa erythropus

Size and description 30cm. Larger with longer legs and bill than Redshank. In breeding plumage uniform black with fine white speckles on wings, in winter pale grey above with white underside and strong eyestripe and supercilium. Adults in spring and autumn are transitional between these two plumages. Juvenile browner with lightly barred underside. Narrow white rump patch shows in flight. Legs red, long bill black with red base, slight 'droop' at tip.
Voice Loud 'chu-it'.
Habitat Passage migrant en route from Arctic to Mediterranean, may overwinter. Shallow freshwater wetlands.
Food and habits Wades, probes and swims energetically in pursuit of aquatic invertebrates. Often seen singly, sometimes small groups.

Redshank

Tringa totanus

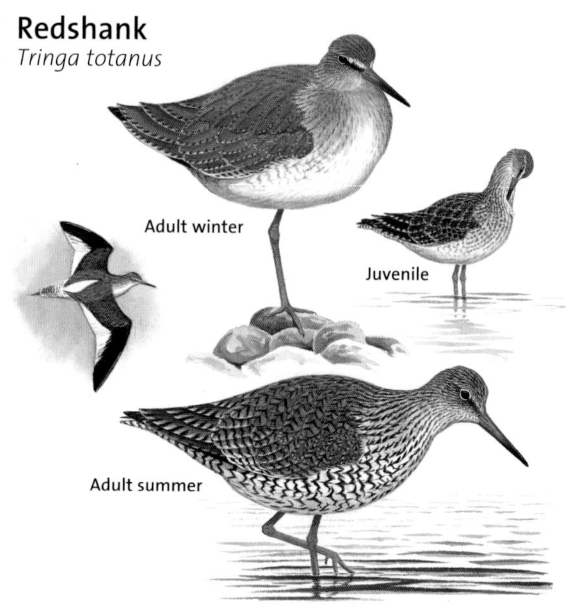

Size and description 28cm. Grey-brown wader with an orange-red bill and legs. Plumage greyer in winter than in summer. White rump and trailing edges to wings noticeable in flight.

Voice Variety of yelping calls. Song, 'tu-udle...', may be given in flight or from the ground.

Habitat Breeds on flood meadows, grassland and lowland moors both near coast and inland. Winters on coast, especially estuaries and mudflats. Most widespread on passage.

Food and habits Feeds mostly on invertebrates. Often perches conspicuously on posts. Nest a deep cup hidden in a grass tussock.

Greenshank

Tringa nebularia

Size and description 32cm. A rather pale, grey wader. Legs are green. In flight shows a white tail, rump and lower back.

Voice Lower pitched call than Redshank. Song, 'ru-tu, ru-tu ...', given in flight or while perched.

Habitat Breeds inland on marsh and moor edges, in northern Europe including a few pairs in Scotland. Widespread passage migrant to coastal marsh, creeks and muddy lake shores.

Food and habits Feeds almost entirely on invertebrates, amphibians and fish. Nests in a scrape well hidden on the ground.

Green Sandpiper
Tringa ochropus

Size and description 22cm. Medium-sized and relatively robust sandpiper with distinctly contrasting plumage. Upperparts very dark, underparts white with a strongly separated dark breast. Juvenile more heavily speckled than adult.
Voice Flight call 'tluit-uit-uit'; warning call 'tip tip'; song 'tloo-i tlui'.
Habitat Mainly passage migrant in northern Europe, including Britain, with a few wintering in southern Britain. Occurs in all types of water body during migration. Breeds in damp forests near fresh water.
Food and habits Feeds on invertebrates, as well as plant fragments. Often nests high in trees in abandoned nests of passerine birds.

Wood Sandpiper
Tringa glareola

Size and description 20cm. Size and plumage similar to that of Green Sandpiper, but longer legged and more delicate. Upperparts mottled brownish-grey with distinct spotting and clear feather edges. Juvenile slightly darker.

Voice A dry 'chiff if if'.

Habitat Rather uncommon passage migrant to Britain, most likely in autumn. Marshland, riverbanks and similar habitats, often near coast but not usually on mudflats.

Food and habits
Wades in shallow water, feeding on invertebrates and plants. Nest a grassy scrape well hidden on the ground.

Juvenile

Adult summer

Common Sandpiper

Actitis hypoleucos

Size and description 20cm. Small wader with brown upperparts, white underparts, white sides to the rump and tail, and a white wingbar.
Voice Call in flight 'twee-wee-wee'. Song more melodious.
Habitat Upland streams and lochs. In non-breeding season occurs on passage on inland waters such as reservoirs and sewage farms, and in coastal areas.
Food and habits Eats mainly invertebrates, and some plant matter. Flicks tail. Nests in a shallow scrape on the ground.

Juvenile

Adult summer

Turnstone

Arenaria interpres

Size and description
23cm. Boldly marked; looks black and white. Short and slightly upturned bill. Rufous markings in breeding plumage give it a tortoiseshell appearance.

Voice Variety of calls; typically short and nasal.

Habitat Breeds on tundra in Arctic regions, extremely widespread in winter on coasts worldwide including all of Britain. Rocky and stony shores favoured, also seaside towns.

Food and habits Diet is mainly insects, molluscs and crustaceans, which it finds by using its bill to overturn pebbles, and pieces of seaweed. Nest a shallow scrape well hidden on the ground.

Red-necked Phalarope

Phalaropus lobatus

Size and description 18cm. Small, dainty wader, usually seen swimming. Breeding plumage dark grey above with buff 'tramlines', white below, with white throat and red neck-sides, female brighter than male. In winter pale grey and white, dark marking behind eye. Juvenile browner, face mostly white. Bill black and needle-like, legs dark, narrow white wingbar shows in flight.

Voice Sharp 'whit'.

Habitat Arctic breeder, migrates to Indian Ocean. Very localised British breeder (Shetland) and uncommon passage migrant. Breeds on small, shallow marshy pools, migrants seen at sea and on coastal wetlands.

Food and habits Takes insects while swimming, uses spinning action to disturb them and pecks rapidly. Often very confiding.

Grey Phalarope

Phalaropus fulicarius

Size and description 23cm. Larger and thicker-billed than Red-necked Phalarope. Breeding plumage mainly brick-red, but in Britain most are juveniles moulting to first-winter plumage, with white head, dark crown and cheek-patch, dark upperparts, pale underparts, and variable amounts of silver-grey plumage appearing on upperside. Later in year, back and wings completely silver-grey.

Voice 'Whit' or 'zit' call.

Habitat Breeds in high Arctic, winters off west Africa. Migrants passing Britain seen mainly offshore along east and south coasts. May retreat to inland lakes and pools to escape stormy weather.

Food and habits Feeds on land more often than Red-necked Phalarope, takes similar prey. Buoyant on the water. Very confiding.

Pomarine Skua

Stercorarius pomarinus

Size and description 46cm. Occurs in a dark and light morph. Juvenile brown and barred. Breeding birds have central tail feathers elongated and twisted by 90 degrees, shaped like spoons.

Voice Alarm call low 'geck'.

Habitat Breeds in tundra. Passage migrant on North Sea and Atlantic coasts, including British coasts.

Food and habits Eats lemmings, eggs and birds in summer, fish in winter, sometimes also stealing or scavenging.

Similar Species **Long-tailed Skua** (*S. longicaudus*, page 114). At 38cm long the smallest skua. Very long tail in breeding birds due to extended central tail feathers. Adult only occurs in a pale morph. Habitat and distribution as for Pomarine Skua.

Arctic Skua

Sterocarius parasiticus

Size and description 40cm. Size of a Common Gull, and the most common of the skuas. Occurs as a dark and pale morph, with brown or whitish underparts, as well as an intermediate variant. More buoyant and graceful flight than that of Great Skua.

Voice Call a meowing 'aag-eeoo'.

Habitat Rare summer visitor and breeding bird on moorland in north Scotland and elsewhere in northern Europe. On migration may be seen at sea from any coast.

Food and habits Summer diet consists mostly of birds, small mammals and insects. Winter diet comprises fish, which are usually taken by piracy from other birds. Nest a grass cup on the ground.

Long-tailed Skua

Stercorarius longicaudus

Adult

Juvenile

Size and description 38cm. Smallest and most elegant skua. Adults occur only in pale morph, light smoky-grey upperside and belly, dark cap, yellow flush on face, pale breast, no obvious white wing flash, central tail feathers slender, pointed and greatly elongated. Juvenile variable, light greyish, dark blackish-brown or intermediate, always cold-toned with strongly barred undertail.

Voice Silent away from nest.

Habitat Arctic breeder, winters in south Atlantic, scarce passage migrant in Britain, usually seen offshore, occasionally resting on beaches.

Food and habits Scavenges and takes living prey opportunistically. Sometimes pirates food from other seabirds. Graceful in flight. Occasionally single adults oversummer in Arctic Skua colonies.

Great Skua

Stercorarius skua

Size and description 54 cm. Large thick-set skua that appears dark brown overall. Sexes are similar, and amount of brown in plumage differs. Large white crescent on upper- and underwing visible in flight.

Voice Call a low 'tok'.

Habitat Breeds in colonies on North Atlantic coasts, and winters at sea. Breeds north Scotland; passage migrant elsewhere in Britain.

Food and habits Harasses other birds into dropping or regurgitating food (klepto-parasitism). Will dive at intruders on its breeding grounds. Nest a bulky grass cup on the ground.

Mediterranean Gull

Larus melanocephalus

Size and description 39cm. Larger and heavier-billed than Black-headed Gull, bill scarlet in summer and winter. Adult pale with white wingtips, dark red legs. In summer has full jet-black hood, white eye-ring, in winter head white with dusky patch behind eye. Juvenile and first-winter have much dark in wings and tail-tip, dark bill and legs, second-years adult-like but retain some black in wingtips.

Voice Loud, interrogative 'kow-wah'.

Habitat Scattered breeding population across southern Europe. Scarce but increasing breeder and winter visitor in southern Britain. Nests on marshland and islands on lakes. In winter on beaches, sometimes inland with gull flocks.

Food and habits Takes live prey and scavenges. Gregarious.

Little Gull

Hydrocoleus minutus

Size and description 26cm. World's smallest gull, compact with rounded wingtips. Adult underwing dark grey, upperwing paler, both with thin white edge. In summer adult's hood solid black, no white eye-ring, white body sometimes with pinkish flush, in winter face white with dark spot behind eye and dusky crown patch. Juvenile extensively scaled blackish on upperside, first-winter has black tail-band and zigzag across wings.

Voice Quiet. Sometimes a repeated 'kek'.

Habitat Breeds north-eastern Europe, winters further south-west. Uncommon passage migrant in Britain, offshore and coastal wetlands, lakes and gravel pits, sometimes in flocks.

Food and habits Invertebrates and small fish. Picks food from surface in graceful, agile, tern-like flight.

Sabine's Gull
Xema sabini

Size and description 39cm. Distinctive small gull with short legs. In all plumages upperside shows broad black wedge at wingtip, grey leading edge of 'hand' and white trailing edge in 'three-triangle' pattern. Adult bill black with yellow tip, summer hood smoky-grey, in winter has dusky neck 'boa' and dark band across head behind eyes. Juvenile scaly grey above, pronounced grey 'boa'.

Voice Single sharp tern-like call, rarely heard.

Habitat Breeds in Arctic North America, passes southern coasts of Britain on migration to African seas, occasionally forced inland by storms.

Food and habits Takes fish and other small marine animals. Graceful tern-like flight. Usually seen singly in Britain.

Ring-billed Gull

Larus delawarensis

Size and description 49cm. Size in between Common and Herring Gull. Has white head and body, mid-grey wings, black white-spotted wingtips. Bill and legs bright yellow in summer, duller in winter (when head also becomes lightly streaked). Eyes pale. Shows an obvious clear-cut black ring near tip of bill. Juvenile mottled grey, becoming more adult-like over successive moults.

Voice Harsh squeals and chatters.

Habitat Rare, mainly winter visitor from North America. In Britain most join flocks of other gulls on the coast.

Food and habits Opportunistic scavenger and predator, like other large gulls highly omnivorous. Same individuals return year after year to preferred winter spots.

Second-winter

Adult winter

Adult summer

Common Gull
Larus canus

Size and description 41cm. Resembles a small Herring Gull, but the legs are yellow-green and the bill lacks a red spot. Grey upperparts, white below, black wingtips and a 'kind' facial expression.
Voice Higher pitched than that of large gulls.
Habitat Coasts; breeds on moorland and freshwater lochs. Mainly resident in Britain, with winter visitors from northern Europe. After breeding many birds move south.
Food and habits Feeds on earthworms, insects, seeds, small mammals, birds and marine invertebrates.

First-winter

Adult summer

Lesser Black-backed Gull
Larus fuscus

Size and description 55cm. Dark grey back, white head and underparts, yellow bill with a red spot, and yellow legs. Juvenile brown, slightly darker than Herring Gull.

Voice Loud calls, deeper than Herring Gull's.

Habitat Breeds on coast and inland in Britain, on wetland and increasingly on buildings. More arrive in winter, widespread on all coasts and inland.

Food and habits Eats almost anything, including fish, small mammals, birds and their eggs, and carrion. May feed at rubbish tips. Nest a bulky mound of flotsam and grass.

Yellow-legged Gull

Larus michahellis

Size and description 62cm. Very like Herring Gull, slightly larger and more robust. Upperparts slightly darker, and legs yellow rather than pink, in winter head becomes less heavily streaked. Juveniles and first-winters paler-headed than same-aged Herring.
Voice Calls include 'kyow' note, deeper than Herring's version.
Habitat Breeds in southern and western Europe. Very rare breeder and regular winter visitor to Britain, mainly in south. Usually among or near other large gull flocks, on the shore, around fishing boats, and inland at reservoir roosts or refuse tips.
Food and habits Very diverse diet includes carrion, fish and some plant material. Will hybridise with Herring and Lesser Black-backed Gulls.

Caspian Gull
Larus cachinnans

Size and description 61cm. Large gull, extremely like Herring and Yellow-legged Gulls. Like the latter, was formerly considered a subspecies of Herring Gull. Wings and back mid-grey, legs dull yellowish-pink. Eye sometimes dark, looks very small, bill looks long, relatively slender and parallel-edged. Head can be very pale in first-winter plumage.
Voice Typical large gull repertoire of squawks, cackles and 'kyow' notes.
Habitat Breeds in south-eastern Europe. Rare autumn and winter visitor, mainly to south. As with Yellow-legged Gull, usually among feeding or roosting assemblages of other large gulls.
Food and habits Scavenges carrion and discarded refuse, also takes fish, small mammals and other live prey.

Herring Gull

Larus argentatus

Size and description 61cm. Silver-grey upperparts, black wingtips, white head and underparts, yellow bill with a red spot, and pink legs. In winter the head and neck are streaked brown.

Voice Wide variety of wailing calls. Loudest is 'kyow-kyow-kyow'.

Habitat Coasts and inshore waters. In Britain and northern Europe abundant on coasts, and common inland in winter. Breeding increasingly inland.

Food and habits Diet includes fish, crustaceans, carrion and birds. Nest a bulky mound of flotsam and grass.

Iceland Gull

Larus glaucoides

Size and description 55cm. Clearly smaller, slimmer and longer-winged than a Herring Gull. Adult has pale grey wings with white tips, otherwise white, with some brown head-streaking in winter, legs pink. First-winter milky tea-coloured with faint streaking, very pale-winged, second-winter almost pure white. Kumlien's Gull, Canadian subspecies, has light grey markings in wingtips.

Voice High-pitched versions of typical large gull calls.

Habitat Arctic (Greenland, Canada) breeder, winter visitor to Britain, especially in north. Usually coastal, often around fishing ports and harbours, sometimes joins other gulls to roost or feed inland.

Food and habits Fish and other animal prey, carrion of all kinds. Usually in flocks with other gulls.

Glaucous Gull

Larus hyperboreus

Size and description 65cm. Plumage almost identical to Iceland at all ages, but is much larger, heavier-built and proportionately shorter-winged. First-winter shows distinctive pink bill with large black tip.
Voice Rather quiet but has the usual gull calls.
Habitat Arctic (including Iceland) breeder, winter visitor to Britain, commoner further north. Usually found on the coast, at spots with rich food supplies such as fishing harbours, occasionally inland with other gulls at refuse tips or on reservoirs.
Food and habits Has the usual diverse gull diet, and when feeding alongside other gulls can dominate and bully all species except Great Black-backed. Will also 'mug' other birds for food.

Great Black-backed Gull

Larus marinus

Size and description 72cm. Very large gull with black upperparts, a white head and underparts, a yellow bill with a red spot, and pink legs. Juvenile is brown.

Voice Deep hoarse calls, 'uk-uk-uk'.

Habitat Resident in Britain. Coasts and islands during breeding season. At other times also on estuaries and inland fresh waters.

Food and habits Eats a wide variety of creatures, including fish, birds, mammals and carrion. May feed at rubbish tips. Nest built from flotsam and seaweed on a ledge.

Kittiwake

Rissa tridactyla

Size and description 41cm. Small gull with a grey back and wings, white head and underparts, dark eye, yellow bill and black legs. Solid black wingtips separate it from Common Gull, which has white 'windows'. Juvenile has a dark 'W' shape across wingspan.

Voice Calls its own name, 'kitt-ee-wayke'.

Habitat Nests on cliffs in northern Europe, in Britain most common in north and north-east, very scattered in south and west. Outside breeding season may be seen offshore anywhere around Britain.

Food and habits Feeds on fish, worms, molluscs and crustaceans. Nests in colonies on cliff ledges.

Black-headed Gull

Chroicocephalus ridibundus

Size and description 37cm. In winter the head is white with a grey-brown crescent behind the eye. Breeding birds have a chocolate-brown head. Bill is red and finer than bills of most other European gulls.

Voice Noisy when in flocks. Calls include a strident 'kee-yah'.

Habitat Nests on coastal marshland and islands in lakes in Britain and northern Europe. In winter very widespread both on coasts and inland.

Food and habits Feeds on seeds and invertebrates, and scavenges in rubbish. Nest a large mound of flotsam and grass erected on the ground.

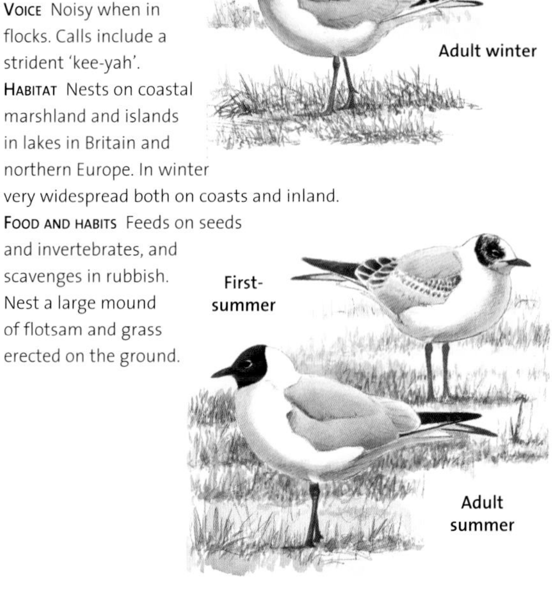

Little Tern
Sternula albifrons

Size and description 23cm. Smallest of the black-capped 'white' terns. The tail is short and barely forked, and the wings are narrow. Breeding adult has a neat white forehead, yellow bill with a black tip, and orange legs. Winter adult and juvenile are duller.
Voice Chattering calls.
Habitat Summer visitor to northern Europe. Rather scarce and scattered in Britain. Nests in colonies on flat sandy and shingly beaches, on migration may be seen offshore from any coast.
Food and habits Diet consists of small fish and invertebrates. Often hovers, employing very rapid wingbeats, before plunge-diving. Nesting birds on beaches are sensitive to disturbance, so decreasing where it is not fully protected.

Black Tern

Chlidonias niger

Size and description 23cm. Small, short-tailed tern. In summer had black body and grey wings and tail. In winter head and body white but with black rear crown and half-collar, first-winter similar but upperside has scaly pattern. Bill black, legs blackish.

Voice Usually quiet, occasionally a sharp 'kik'.

Habitat Breeds patchily across western Europe, migrates to Africa. Passage migrant in Britain, often offshore but also inland where it may linger to feed over well-vegetated marshy pools and lakes.

Food and habits Feeds mainly on aquatic insects, with small fish more important in winter. Most prey picked from water's surface in graceful dipping and turning flight. Often in small parties.

Sandwich Tern

Sterna sandvicensis

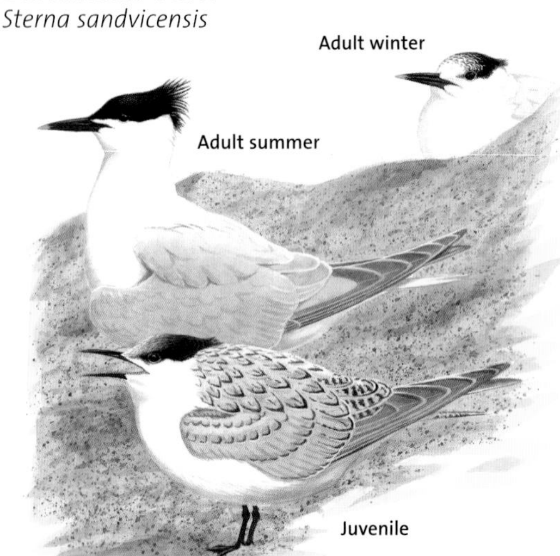

Size and description 41cm. Large tern with pale plumage, rather short black legs, and a long and slender black bill. Breeding adults have a black crown with a shaggy crest, and a yellow tip to the bill. White forehead in winter.

Voice Distinctive harsh 'kirrick'.

Habitat Summer visitor to northern Europe. Fairly common breeding bird in Britain, forms large colonies on beaches and islands on coastal lagoons. On migration may be seen offshore from any coast. Occasionally overwinters.

Food and habits Eats fish and other marine invertebrates. Nest a simple scrape in sand.

Common Tern

Sterna hirundo

Size and description 35cm. Grey upperparts, a black crown, and dark red legs and bill, which has a black tip. Long forked tail. White forehead in winter.

Voice Call is a strident 'keeyah' and 'wik-kik-kik'.

Habitat Summer visitor to Britain, fairly common and widespread on coasts and inland. Nests on beaches and lake islands, fishes over sea and fresh water. On migration may be seen offshore from any coast.

Food and habits Eats fish, worms, insects, molluscs and crustaceans. Often dives for fish. Nests on dunes, salt marshes and shingle banks, in colonies or as single pairs.

Roseate Tern

Sterna dougallii

Size and description 38cm. Whiter than Common or Arctic Tern, stands a little taller and has longer tail streamers. Bill long, black with hint of red at base. Underside shows subtle rosy flush. Juvenile heavily marked dark grey on upperside with black cap, forehead whitens in first-winter plumage.

Voice A grating 'kraaak' and disyllabic sharp 'ke-wick'.

Habitat Very widespread globally but rare in Britain, most colonies in Ireland. Migrates to Africa in winter. Nests on undisturbed rocky coasts, especially on small islands.

Food and habits Catches small fish especially sand-eels. Has a stiff-winged, strong flight and will plunge-dive from a greater height than Common or Arctic Tern.

Arctic Tern

Sterna paradisaea

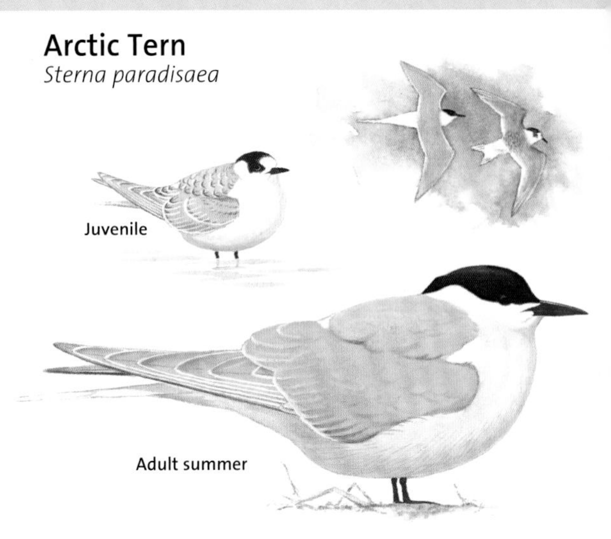

Size and description 35cm. Breeds mostly by coast in northern Britain. Elsewhere generally seen on passage. Similar to Common Tern, but with a blood-red bill. In winter has a white forehead with a black bill. Tends to have longer tail streamers.

Voice Calls similar to those of Common Tern; also a high whistling 'kee, kee'.

Habitat Summer visitor to northern Europe, in Britain very rare in south but has some huge colonies in northern England and Scotland. Nests on stony beaches and lake islands. On migration may be seen offshore from any coast.

Food and habits Eats fish, insects, molluscs and crustaceans. Nest a shallow scrape in grass or sand.

Common Guillemot

Uria aalge

Size and description 43cm. The most common auk in Europe. Has short stubby wings, pied plumage and short legs. Upperparts are black/brown. Some birds have a white line over the eye and are known as 'bridled'. In winter the throat is white.

Voice Call is a caw, 'aargh'.

Habitat Breeds in dense colonies on sheer sea cliffs in northern Europe, in Britain commoner in north, does not breed in south-east. In winter may be seen offshore from any coast, occasionally ill or injured birds are found on beaches.

Food and habits Eats fish and other marine animals. Lays eggs on narrow cliff ledges. Colonies can comprise many thousands of pairs.

Razorbill

Alca torda

Size and description 41cm. A black-and-white bird with a strong vertically flattened black bill. Juvenile and non-breeding birds have a white throat.

Voice Makes a whirring sound and growls.

Habitat Breeds on sea cliffs in northern Europe, in Britain mainly in north, scarcer and uses wider ledges than Guillemot. In winter may be seen offshore from any coast, especially sheltered bays and harbours.

Food and habits Diet almost entirely marine creatures. Fish are caught by diving from the water's surface and pursuing prey underwater by flapping the wings like flippers. Flight fast and whirring, usually low over the water. Breeds in colonies on rocky coasts with cliffs.

Black Guillemot

Cepphus grylle

Size and description 35cm. Smaller and squatter than Guillemot. In breeding plumage black with white circular marking on wing (upper and underside). In winter, mostly pure white, with same black-and-white wing pattern, and black scaling on back, rump and crown. Scarlet feet and gape consicuous all year round.

Voice A thin whistle given at breeding sites.

Habitat Breeds around northern and north-west Europe including north-west Britain, and the Arctic. Nests on rocky coasts and man-made structures, most overwinter at sea near breeding site.

Food and habits Swims and dives to catch fish. Nests colonially, performs communal displays. Rarely associates with other auks.

Little Auk

Alle alle

Adult winter

Size and description 20cm. Tiny, dumpy auk with very short, stout bill. Wings proportionately much longer than larger auks. In breeding plumage has white belly and wingbar, otherwise blackish, in winter breast, chin and cheek-sides become white.

Voice Silent away from breeding grounds.

Habitat Breeds in high Arctic, wanders southwards in winter. Passage migrant to Britain, passing mainly east coasts in late autumn, usually scarce but sometimes in large numbers.

Food and habits Dives from surface for plankton and other small marine life. Takes off and manoeuvres in flight much more easily than larger auks. After autumn gales may be 'wrecked' on land.

Puffin

Fratercula arctica

Size and description 30cm. Instantly recognizable by the bright bill and clown-like face markings. Smaller than Razorbill and Guillemot. In winter the bill is smaller and greyer, and the face is smudged.
Voice Growling 'aar'.
Habitat Nests in burrows on grassy clifftops in north-west Europe, in Britain mainly in north. Winters well offshore, not often seen from coasts.
Food and habits Eats mainly fish, especially sand-eels, and capable of holding several at a time in its notched bill. Highly gregarious, nesting in huge colonies. Nests in burrows or rocky crevices.

Feral Pigeon *Columba livia* var. *domestica*

Rock Dove *Columba livia*

Size and description 33cm. With black wingbars and a white rump, many feral pigeons resemble the Rock Doves from which they originate. However, colours vary from white to very dark grey, and some may be pale fawn.

Voice A soft cooing.

Habitat Feral pigeon common in Europe in towns and cities, where it breeds on buildings. Rock Dove locally common in mountains and on rocky coasts, particularly in southern Europe. In Britain local on coasts in western Scotland and western Ireland.

Food and habits Seeds, grain and discarded human food. Rock Dove nests in hollows and crevices, and on rock ledges.

Cuckoo

Cuculus canorus

Size and description 34cm. Slim-looking bird with long swept-back wings and a long rounded tail. Grey with a pale, barred breast. Some females may be rufous. Juvenile barred, brown, with a white patch on the nape.

Voice Male gives well-known 'cuckoo' call; female has a bubbling trill.

Habitat Widespread though not especially common summer visitor to Britain, uses all habitat types where its host species occur, including coastal marsh and uplands.

Food and habits Eats insects and is capable of swallowing hairy caterpillars. Females lay eggs in other birds' nests.

Barn Owl

Tyto alba

Size and description 34cm. Golden-spangled back, heart-shaped pale face and white underparts. Longer wings and legs than Tawny Owl.
Voice Call is a screech; also makes hissing and snoring sounds.
Habitat Resident in Britain. Fields, meadows and marshes; needs open country with rough grassland for hunting.
Food and habits Feeds mainly on rodents, especially rats and voles. Often nests in buildings, usually old barns and farm outhouses. Largely nocturnal, and often seen in car headlights as it searches verges for prey.

Long-eared Owl

Asio otus

Size and description 34cm. Usually looks slender though this varies with alertness. Cryptic mottled brown plumage with orange tones to breast and face, eyes orange, ear-tufts long (though may be held flattened). Underpart streaking covers entire belly. Wings long, with orange patch on upperside primary bases.

Voice Low repeated hoots form territorial song. Begging chicks have 'squeaky-gate' call.

Habitat Breeds across much of northern hemisphere. Uncommon in Britain though some European birds move to Britain in winter. Light woodland and woodland edges, hunts over more open ground.

Food and habits Preys on small rodents. Hunts at night, by day roosts in thick cover, sometimes in small groups.

Short-eared Owl

Asio flammeus

Size and description 37cm. Larger than Long-eared Owl, paler with colder, yellower plumage tones, streaking on underside fades leaving unmarked pale belly. Eyes yellow with black surround giving intense expression. Shows large yellowish patch on primary bases of upperwing. Ear-tufts short and inconspicuous.

Voice Barking call, hollow series of hoots when breeding.

Habitat Very wide global distribution. Nomadic habits, moving according to prey availability. Uncommon breeder in upland Britain, more numerous and coastal in winter. Moors, heaths and, especially in winter, rough grassland.

Food and habits Feeds mainly on Short-tailed Voles. Hunts by day as well as dusk and dawn, with low quartering flight. Sometimes several patrol the same field.

Adult

Common Swift

Apus apus

Adult

Size and description 17cm. Long and narrow crescent-shaped wings, a torpedo-shaped body, a short forked tail and very short legs. Plumage is dark brown with a pale throat.

Voice Shrill monotone scream, often uttered by tight flocks flying around buildings at roof-top height.

Habitat Summer visitor to northern Europe, breeds throughout Britain, mainly on buildings in villages and towns. Large numbers feed over wetlands, including on the coast, particularly just before and after the breeding season.

Food and habits Adapted to feed on high-flying insects, which it catches in its wide gaping mouth. Shuffles around nest site on short legs. Most of its life is spent on the wing.

Juvenile

Kingfisher
Alcedo atthis

Size and description 18cm. Although brightly coloured, Kingfishers are well camouflaged when perched among leaves. Bill is black, but female has a reddish base to lower mandible. Juvenile has a pale spot at the tip of its bill.
Voice Distinctive whistle, 'tee-eee' and 'tsee'.
Habitat Resident in Britain. Rivers, streams and lakes. Visits garden ponds to take small ornamental fish. In winter more coastal, especially if there is a freeze inland.
Food and habits Fish are the main food. Hunts by diving into water from a perch, or by hovering and then diving. Excavates breeding tunnels in steep sandbanks.

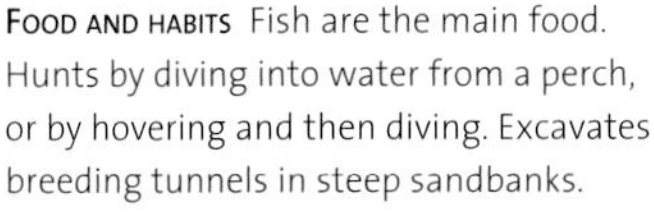

Male

Wryneck
Jynx torquilla

Size and description 17cm. Small, peculiar member of the woodpecker family, with cryptic brown and grey plumage patterned to resemble tree bark. Short bill, rather long tail, and habit of raising crown feathers and twisting the neck.

Voice Has ringing 'quee-quee-quee' song but usually silent in Britain.

Habitat Summer visitor to most of Europe but in Britain only as a scarce passage migrant, mainly in scrub or light woodland along east coasts in autumn, following easterly winds.

Food and habits Feeds mainly on ants but also takes other insects. Feeds on ground, moving with slow hopping gait. Migrants can sometimes be very approachable.

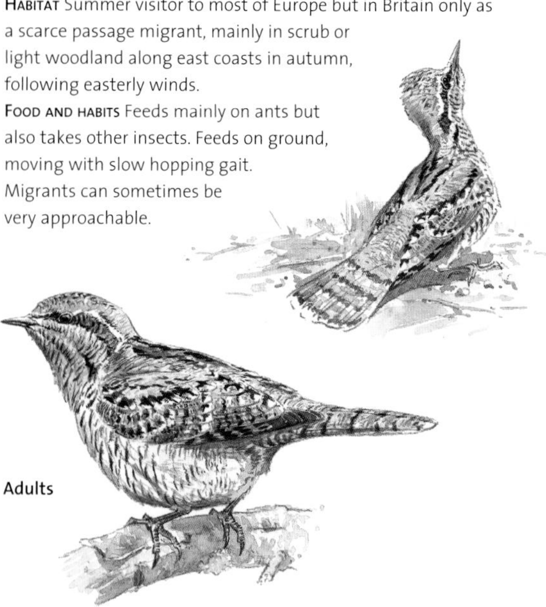

Adults

Green Woodpecker

Picus viridis

Size and description 33cm. Green plumage, but with a distinctive yellow rump and a red cap. Juvenile is speckled and appears more grey. A pale eye and black face and moustachial stripe give the bird a 'fierce' appearance. Male has a red centre to his moustachial stripe, while female's is black. Flight is deeply undulating.

Voice An unmistakable shrill laughing call. Rarely drums.

Habitat Resident in Britain and western Europe. Nests in tree holes but forages in open habitats including coastal meadows where there are plentiful anthills.

Food and habits Feeds on insect grubs and ants, for which it probes soil and rotten wood. Often seen feeding on large open areas of grass. Nests in a hole made in a tree.

Skylark

Alauda arvensis

Size and description 18cm. Streaked brown upperparts, short crest not always obvious, white outer tail feathers. Walks rather than hops. Towering and hovering song flight.

Voice Lengthy warbling song delivered in flight as bird rises vertically, then drops through the air.

Habitat Resident in Britain and western Europe. Breeds on grassland and farmland. Some upland breeders move to the coast in winter and may forage on beaches.

Food and habits Eats insects, worms and seeds. Nests on the ground. Flocks in winter, when numbers are swollen by European migrants. Nest a grassy cup well hidden on the ground. Common but declining.

Shore Lark

Eremophila alpestris

Size and description 17cm. Attractive Skylark-sized bird with light sandy grey-brown upperparts, white belly, and yellowish face with darker crown, bold black eye-mask and breast stripe. In breeding plumage male head adorned with pair of short tufts or 'horns' on rear crown.
Voice Call is a high-pitched 'see-tsi'.
Habitat Breeds in the Arctic, moving south in winter. Occasionally attempts to breed in upland Scotland. Primarily in Britain a passage migrant and winter visitor, most often found along quiet stretches of shingle or sandy beach.
Food and habits Feeds on seeds and small invertebrates, may search the strandline in hunched, shuffling stance. Will flock with Skylarks and buntings.

Sand Martin

Riparia riparia

Adults

Size and description 12cm. Small brown bird with white underparts, a brown breast-band and a short forked tail.

Voice Twittering song, not as musical as that of Swallow.

Habitat Summer visitor to western Europe, in Britain very widespread but patchy, because of precise breeding habitat needs. Nests in stable steep banks, sandy or muddy, including coastal undercliffs, often hunts insects over lakes.

Food and habits Eats insects such as midges caught in flight. Nests colonially in burrows excavated in sandbanks. On migration may roost in large numbers in reed beds.

Swallow

Hirundo rustica

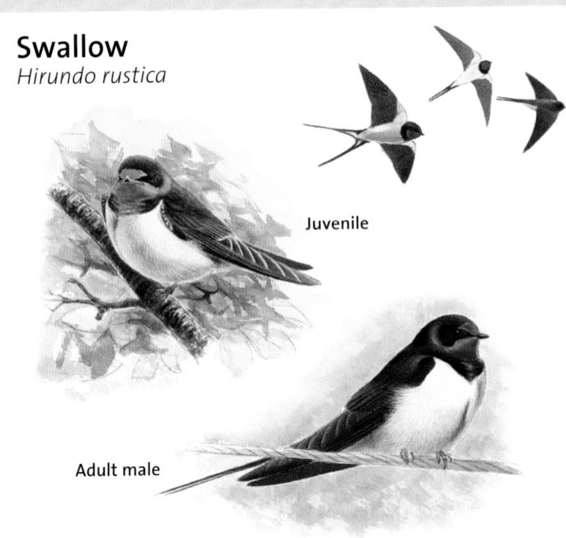

Size and description 19cm including tail of 3–6.5cm. Wings long and pointed, tail deeply forked. Pale cream underparts, dark blue wings and back, and a red throat with a blue-black neck band. Fast flight with powerful wingbeats.

Voice High-pitched 'vit-vit' call in flight. Warning call for cats and other ground predators a sharp 'sifflit'; for birds of prey, 'flitt-flitt'. Song a rapid rattling twitter.

Habitat Summer visitor to western Europe, in Britain common and widespread, nests mainly in buildings, feeds over open countryside. Flocks often seen moving along coast in autumn, or gathering to roost in coastal reedbeds.

Food and habits Feeds on insects, which it catches in flight by flying low over fields and water. Cup-shaped clay nest built in buildings.

House Martin
Delichon urbicum

Size and description
14cm. Wings broader than Swallow's and forked tail shorter, giving a stubbier appearance. Rump white, and wings, head and tail dark blue. Flight more fluttery than Swallow's, with flaps often interspersed with glides. Underparts of juvenile usually duskier white than adult's.
Voice Harsh twitter. Song a series of formless chirps.
Habitat Summer visitor to western Europe, in Britain common and widespread, nests on buildings and in some areas sea cliffs, feeds over open countryside. Pre-migratory gatherings seen on the coast in autumn.
Food and habits Tends to feed on flying insects at greater altitude than Swallow. Rarely on the ground, except when collecting mud for nest. Builds rounded mud nest under protrusions on buildings, and sometimes cliffs.

Meadow Pipit

Anthus pratensis

Size and description 15cm. Streaked brown upperparts; underparts spotted. Darker legs than Tree Pipit. Best identified by call.

Voice Call 'pheet' uttered 1–5 times. Song given from perch or in display flight as it describes an arc from the ground.

Habitat Mainly resident in Britain and western Europe. Breeds on rough grassland including cliff-tops, some upland birds move to coastal areas in winter.

Food and habits Eats mostly insects; also spiders, earthworms and some seeds. Nests on the ground in a small depression.

Rock Pipit

Anthus petrosus

Size and description 16cm. Larger than Meadow Pipit. Plumage mostly drab dusky grey-brown, paler below but with heavy diffuse grey streaking, strong pale eye-ring, legs dark, yellowish base to dark bill. In flight shows greyish outer tail feathers.

Voice Call a single strong 'pseep'. Song, given in flight, long series of accelerating call notes.

Habitat Occurs on rocky coasts around north-western Europe, most northerly populations move south in winter. Most coastal of all songbirds, rarely found inland. Nests in rock crevices, also sometimes buildings in seaside towns and villages.

Food and habits Feeds on small invertebrates. Runs or walks, pauses in upright stance on prominent rocks. Often quite confiding.

Water Pipit

Anthus spinoletta

Size and description 16cm.
Very similar to Rock Pipit but paler with strong whitish supercilium, white outer tail feathers. In winter plumage pale underside is strongly streaked, in breeding plumage it becomes less streaked and develops variable peachy flush.

Voice Similar to Meadow Pipit but slightly stronger and louder. Rarely sings in Britain.

Habitat Has patchy, complex European distribution, many populations making local movements in winter. Winter visitor only in Britain, mainly in south. Freshwater marshland and wet meadows.

Food and habits Feeds on insects and small invertebrates of all kinds, found by searching on the ground. Sometimes in small flocks. Much less approachable than Rock Pipit.

Yellow and Blue-headed Wagtails

Motacilla flava

Size and description 16cm. Several subspecies, with Yellow, *M. f. flavissima*, by far the most common in Britain. Head green with a yellow throat and supercilium; mantle a brighter yellow-green; slender black legs. Blue-headed *M. f. flava* male has a pale blue head.
Voice Call a rich 'tseep'. Song a simple scratching 'sri'srit sri...'
Habitat Meadows, farmland and marshes. *M. f. flavissima* breeds in Britain and on neighbouring European coasts from France to Norway. *M. f. flava* occurs on much of the Continent. Winters in Africa.
Food and habits Insectivorous. Grassy cup nest well concealed on the ground. In decline since the 1980s, probably due to loss of habitat.

Grey Wagtail

Motacilla cinerea

Size and description 19cm. Longest tailed of European wagtails. Grey above and lemon yellow below, with colour particularly strong under the tail, and pink legs. Summer adult male has a distinctive black throat. Tail is constantly wagging.

Voice Call a sharp 'tzit'. Song a simple and metallic 'ziss-ziss-ziss'.

Habitat Habitat Resident in Britain and Europe except far north. Breeds and forages by running water, more widespread in winter including on beaches and in towns.

Food and habits Insectivorous; often chases insects over the water. Nest a grassy cup usually hidden in a cavity near water.

Pied and White Wagtails

Motacilla alba

Size and description 18cm. Male of British race (*M. a. yarrellii*) has a black back and wings, female a dark grey back. In continental race (*M. a. alba*), both male and female have a pale grey back.

Voice Flight call a 'chissick'. Song plain and twittery.

Habitat Pied Wagtail a widespread, common resident in Britain, joined by migrant White Wagtails in autumn. Open habitats of all kinds, often on beaches and rocky coasts in winter.

Food and habits Runs rapidly after flying insects. On the ground its gait is rapid, and its head is moved backwards and forwards while wagging its tail. Prefers feeding on lawns and roofs, and in car parks and roads, where prey is easily spotted. In winter roosts in large flocks.

Black Redstart

Pheonicurus ochruros

Size and description 14cm. Darker than Common Redstart. Breeding male slaty-black above with a black face and breast, and a white flash in the wing. Female duller brown. Reddish rump and tail.

Voice Call a quiet 'tsip-tsip'. Song a short high-pitched warble punctuated by characteristic gravelly notes.

Habitat Rare breeding bird in Britain, mainly south-east, more widespread in winter, much commoner on near continent. Cliffs, rocky coasts and very urbanised areas.

Food and habits Eats mainly insects. Constantly shivers tail. Often nests in wall cavities.

Common Redstart

Phoenicurus phoenicurus

Size and description 14cm. Male has a grey back, black face and throat, white forehead, and bright chestnut breast and tail. Female is a duller brown. Tail is waved up and down.

Voice Calls 'hooeet' and 'kwee-tucc-tucc'. Song a squeaky warble.

Habitat Widespread summer visitor in northern Europe, in Britain more common in north. Breeds in deciduous woodland, on migration often lingers in scrubby coastal areas.

Food and habits Eats mainly insects; also worms, spiders and berries. Nests in a tree hollow.

Stonechat

Saxicola torquatus

Size and description 12cm. Male has a black head, white patch on sides of neck, white wingpatch and dark brown upperparts. Female duller with streaked brown upperparts.

Voice Call a persistent 'tsak, tsak', like two stones being hit together; plaintive song.

Habitat Resident in Britain and much of Europe. Nests in heathland and similar scrubby habitats, some upland breeders move to the coast in winter.

Food and habits Diet is chiefly insects; also worms and spiders. Nests on the ground, often under the cover of a gorse bush.

Juvenile

Female

Male summer

Wheatear

Oenanthe oenanthe

Size and description 15cm. Breeding male has a blue-grey back and black eye-mask, wings and lower tail; distinctive square white rump and upper tail visible in flight. Winter male is browner.

Voice Call 'chack, chack'.

Habitat Summer visitor to mainly northern Britain and north-west Europe. Breeds mainly on rocky uplands, during migration found on beaches and grassy areas on the coast.

Food and habits Eats mostly insects. Nests on the ground in rabbit burrows, holes under stones and stone walls.

Ring Ouzel

Turdus torquatus

Size and description 24cm. Dull plumage, sooty black in male, sooty brown in female. White crescentic patch in male, often obscure in female and juvenile.

Voice Call an excitable 'tack tack'. Song melodious, with 2–4 repeated flute-like tones.

Habitat Fairly uncommon summer visitor to remote uplands in Britain, also north-west Europe. Breeds mainly on rocky uplands. On migration occurs in scrubby areas and fields on the coast.

Food and habits Omnivorous, consuming a wide range of insects, rodents, lizards and berries. Nest a neat cup in bushes or among rocks.

Cetti's Warbler

Cettia cetti

Size and description 14cm. Sturdy dark warbler with very short wings and long, heavy, round-ended tail. Plumage warm red-brown, mid-greyish below, dark eyestripe and pale supercilium. Legs dull pinkish.

Voice Song an abrupt burst of powerful, fluty notes, heard all year round.

Habitat Found across most of southern Europe, recent colonist of southern Britain. Resident. Occurs in damp and well-vegetated areas including reedbed edges and overgrown ditches.

Food and habits Feeds on all manner of small insects and other invertebrates. Extremely skulking and difficult to observe, slipping mouse-like through thick, low vegetation. As a non-migrating insectivore, can suffer population crashes in very severe winters.

Grasshopper Warbler

Locustella naevia

Size and description 12.5cm. Small, rather drab warbler with long, round-ended tail. Plumage dull grey-brown, paler below, throat whitish. Darker streaking on crown, back and on the very elongated undertail-coverts.

Voice Song a continuous extremely fast dry trill, sounding like an angler's reel. Call a crisp 'stit'.

Habitat Summer visitor to most of Europe, widespread but rather uncommon in much of Britain. Long weedy grass and marshland edges.

Food and habits Insect-eater. Usually difficult to see, its strange song hard to pin down, and often sings from within deep cover, but some individuals more showy. Moves low and mouse-like through vegetation.

Sedge Warbler

Acrocephalus schoenobaenus

Size and description 13cm. Olive-brown streaked-backed warbler with a rounded tail and rufous-coloured rump. Conspicuous creamy-white stripe above the eye. Sexes look similar.

Voice Loud, jumbly and scratchy song.

Habitat Waterside vegetation near reed beds, rivers and lakes, and lowland marshes; also dry scrubby areas. Summer visitor to Britain, migrating to Africa in late summer.

Food and habits Mainly eats insects; takes berries in autumn. Nests in rank vegetation.

Reed Warbler

Acrocephalus scirpaceus

Size and description 13cm. Small olive-brown warbler with slight rufous tinge to its upperparts. Buff-coloured below. Rounded tail. Sexes look similar.

Voice Monotonous churring song.

Habitat Mainly reed beds. Summer visitor to Britain.

Food and habits Eats water insects. In autumn feeds on berries, which provide energy for its long migratory flight. Builds a nest of woven grasses slung between reed stems. Common host to Cuckoo.

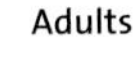

Adults

Common Whitethroat

Sylvia communis

Size and description 14cm. Male has a grey head, a bright white throat, brown upperparts and pale underparts. Female has a brown head. Tail long and slim.

Voice Call a sharp 'tacc, tacc'. Song a rapid warble.

Habitat Open woodland, gardens, hedgerows and scrub. Summer visitor to Britain.

Food and habits Eats mainly insects, and some fruits and berries in autumn. Nests in brambles and low bushes not far from the ground.

Common Chiffchaff

Phylloscopus collybita

Adults

Size and description 11cm. Small neat bird with a fine bill and thin legs. Very similar to Willow Warbler. Primaries shorter. Legs usually dark and bill even finer. Stripe above the eye less distinct, while darkish patch beneath the eye emphasizes white eyering.

Voice Call a soft 'hueet'. Song a distinctively slow 'chiff-chaff-chiff-chaff'.

Habitat Open deciduous woodland with some scrub. A few overwinter, often in sheltered coastal areas.

Food and habits Similar to Willow Warbler.

Willow Warbler

Phylloscopus trochilus

Size and description 11.5cm. Head, back and tail generally brownish-green; throat and eyebrow yellowish; legs usually pale brown. Primary feathers project beyond tertials.

Voice Call a soft 'huitt' similar to Common Chiffchaff's. Song rather sad.

Habitat Summer visitor to northern Europe, common and widespread in Britain, especially the north. Breeds in mainly deciduous woodland, on migration visits coastal copses and scrub.

Food and habits Feeds on insects found among leaves. Nest a grassy dome on or near the ground.

Yellow-browed Warbler
Phylloscopus inornatus

Size and description 10cm. Tiny, Goldcrest-like warbler with green and yellow plumage. Has bold face pattern with green crown, yellow supercilium and blackish eyestripe, wings have double yellow wingbar, rest of upperside bright moss-green, underside whitish.
Voice Disyllabic high-pitched 'soo-eet'.
Habitat Breeds in Siberia, occurs in Britain as a vagrant or rare passage migrant. Most appear in mid-autumn on the east coast, often in small stands of sycamore trees (these keep their leaves longer into autumn and therefore hold more insect prey).
Food and habits Searches leaves for tiny insects, often hovers, extremely active. May join tit flocks. Occasionally overwinters.

Pied Flycatcher
Ficedula hypoleuca

Size and description 13cm. Breeding male has bold black-and-white plumage. Shorter tailed and more compact than Spotted Flycatcher, with a white wingbar. Constantly flicks its wings and tail.

Voice Calls include a metallic 'whit'. Song quite shrill, 'zee-it, zee-it', interspersed with trills.

Habitat Summer visitor to northern Europe, in Britain commoner further north. Breeds in deciduous woodland, on migration occurs in coastal scrub.

Food and habits Feeds on insects caught on the wing, and sometimes on the ground. Seldom hunts from the same perch twice. Breeds in hollows and may use nestboxes.

Bearded Tit

Panurus biarmicus

Size and description 15cm. Tit-like with a plump body and very long and broad tail. Both sexes have rich orange-brown plumage, and male also has a grey head and black moustaches.

Voice Call like tiny bells, a ringing 'ching ching'. Song a softly chirping 'tship tship tshir'.

Habitat Resident in western Europe. Favours reed beds. In Britain quite common in south-east England.

Food and habits Feeds on insects, and reed seeds in winter. Shuffles up and down reed stems. Builds an open nest of stems in reed beds.

Chough
Pyrrhocorax pyrrhocorax

Size and description 39cm. Elegant smallish crow with glossy black plumage, red legs and long, downcurved red bill. Juvenile has shorter, more yellowish bill. In flight shows strikingly long separated primary feathers ('fingers').
Voice A Jackdaw-like but softer 'kyaa'.
Habitat Mainly found in southern Europe. In Britain exclusively coastal, and found only in western Ireland, Isle of Man, Wales, west Scotland and the tip of south-west Cornwall. Nests in hollows on sea cliffs, forages on coastal fields, grassy clifftops and beaches.
Food and habits Mainly invertebrates, especially worms and other soil-dwelling animals. Sociable, usually seen in small groups. An aerial expert, often 'plays' on air currents.

Jackdaw

Corvus monedula

Size and description 33cm. Nape is grey and eye has a very pale iris. In flight, wingbeats are faster and deeper than Carrion Crow's. Struts as it walks. Flies in flocks almost as densely as pigeons.

Voice Calls a metallic high-pitched 'kya' and 'chak'.

Habitat Common resident in Britain and much of Europe. Nests in towns, quarries and on sea cliffs, forages in all kinds of open habitat including town streets and on beaches.

Food and habits Feeds on invertebrates, eggs, nestlings and grains. Breeds in tree hollows or on ledges of buildings and cliffs, in pairs or small colonies.

Carrion Crow

Corvus corone

Size and description 47cm. Totally black with a stout bill. Upper leg feathers neatly close-fitting. Juvenile much like adult, but duller.
Voice Call a croaking 'krra-kra-kraa'.
Habitat Resident in Britain. A wide variety of habitats, from coast to mountains and towns, throughout western and central Europe.
Food and habits Omnivorous; feeds on carrion, nestlings and eggs, grain and insects. Not colonial. Nest a bulky twig structure high up in tree canopy.

Hooded Crow

Corvus cornix

Size and description 47cm. Same size and build as Carrion Crow, but black restricted to head, breast centre, wings and tail, otherwise dull grey-brown. Individuals with more extensive black are likely to be Hooded x Carrion Crow hybrids – such birds common in areas where both species occur.

Voice Harsh caws and croaks.

Habitat Occurs over northern and central Europe, and Ireland and north-west Scotland. No special habitat needs, found in woodland, farmland, moor, heath and coast. Nests in trees or on cliff ledges.

Food and habits Very diverse diet of almost anything animal and vegetable. Mainly forages on the ground. Quite gregarious. Quick to mob passing birds of prey.

Raven

Corvus corax

Adult

Size and description 61cm. Largest crow and largest passerine. Heavy head has shaggy throat feathers and a huge bill. In flight shows broad heavily fingered wings, a protruding head and a wedge-shaped tail.

Voice Calls deep and croaking 'korrrk', 'klong' and repetitive 'korrp korrp korrp'.

Habitat Coastal, forest and mountain areas year-round in much of Europe, though largely absent from central Europe. In Britain found only in west and Ireland.

Food and habits Feeds on carrion, as well as small mammals, birds, molluscs and vegetable matter. Builds a bulky nest from twigs in a tree or on a rocky ledge.

Starling
Sturnus vulgaris

Size and description 21cm. Short tail and neck, upright stance, pink legs, white spots and metallic green shine. Non-breeding plumage has clear pale spots, which are reduced in breeding male. Breeding male also has a yellow bill; bill otherwise blackish. In flight, has an arrowhead profile. Flocks fly in tight formation. Juvenile grey-brown.
Voice Versatile mimic of other birds. Calls are creaky twitters, chirps, clicks and whistles.
Habitat Widespread throughout Europe in all habitats, particularly human settlements. In Britain large flocks roost in coastal reedbeds in winter.
Food and habits Eats berries, seeds and fruits. Breeds in holes. Outside breeding season roosts in huge flocks in city buildings and trees.

Goldfinch
Carduelis carduelis

Size and description 13cm. Red face, white cheeks and throat, black cap and black-and-gold wings. In flight wings show broad golden bands, and white rump and black tail are visible. Sexes alike, but juvenile has a brown-streaked head.

Voice Cheerful trisyllabic 'tickelitt' call. Song a series of rapid trills and twitters.

Habitat Resident in Britain. Open lowland woodland, heaths, orchards and gardens in most of Europe. Flocks often found around coastal fields and scrub in winter.

Food and habits Eats seeds and berries. Favours teasels and thistle heads. Nest made of hair and rootlets; positioned high in canopy.

Linnet

Carduelis cannabina

Size and description 13cm. Breeding male has a crimson forehead and breast, and a chestnut mantle. Winter male resembles female.

Voice Canary-like song is a pleasant twitter consisting of chirping and rolling sounds, sung from the top of a bush.

Habitat Resident in Britain. Open fields with bushes and waste ground. Farmland and coasts in winter. Widespread throughout most of Europe.

Food and habits Mostly eats seeds and arable weeds. Often breeds in loose colonies. Nest a grassy cup well hidden in a shrub. Common but declining due to changes in agricultural practices.

Twite

Carduelis flavirostris

Size and description 13cm. Small, rather drab brown, small-billed, streaky finch, similar to female Linnet. Male has variable pink patch on rump, and both sexes a marked cinnamon-brown flush on the face and upper breast. Grey bill becomes yellow in winter.

Voice Various rather harsh, nasal calls. Song a simple combination of call notes.

Habitat Breeds on coasts and uplands in northern Europe including northern Britain, moves south in winter. Prefers moorland edges, upland farmland and, in winter, coastal fields and partly vegetated upper slopes of beaches.

Food and habits Feeds on seeds, and in summer small invertebrates. Flocks in winter. Mainly feeds at ground level.

Lapland Bunting
Calcarius lapponicus

Adult winter

Size and description 15cm. Plump, streaky bunting with bold face pattern and strong rufous tones on nape. Bill yellow with dark tip. Breeding-plumaged male (rarely seen in Britain) has black face and breast with white supercilium joining up with white breast sides. Legs black, hind claw elongated.
Voice In flight gives a dry, rattly 'terrreck'.
Habitat Arctic breeder, migrating south and west in winter, uncommon winter visitor to mainly east coasts in Britain. Often in stubble fields or weedy set-aside, or on quiet beaches.
Food and habits Feeds on seeds in winter. Forages on ground in low, shuffling gait. Often flocks with Snow Buntings.

Snow Bunting
Plectrophenax nivalis

Size and description 17cm. Large, long-winged, mostly white bunting. Breeding male white with black back, wings and tail, large white wing patch, black legs and bill. Female grey above, white below. In winter both sexes mostly white but with gingery brown crown, cheek-patch and neck sides, tail and primaries blackish, mantle scaled brown, bill yellow with dark tip.

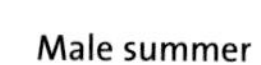

Male summer

Voice Flight call a sweet, rolling twitter.

Habitat Arctic breeder, a few pairs breed in mountainous parts of north Scotland. In winter moves to coasts, more numerous further north. Forages on beaches and fields.

Food and habits Feeds on seeds in winter. Gregarious, methodical slow feeder, extremely approachable.north and east.

Male winter

Corn Bunting
Emberiza calandra

Size and description 18cm. Large plain bunting that is brown with dark streaks. No white markings.

Voice Distinctive monotonous jangling song is high-pitched, likened to keys being jingled.

Habitat Resident in Britain, common in some areas but with patchy distribution, widespread in Europe. Mainly on farmland, both inland and coastal.

Food and habits Mostly eats seeds, corn, fruits and other vegetable matter; also insects and earthworms. Very sedentary. Sings from a prominent perch. Nest a grassy cup well hidden on the ground.

Cirl Bunting

Emberiza cirlus

Size and description 15 cm. Slightly smaller than Yellowhammer. Male has yellow face with blackish crown, eyestripe and chin, all connecting to dark green nape patch. Underside yellow with green breast-band shading into rufous flanks, upperside streaky and rufous. Female greyer than female Yellowhammer.

Voice Song like Yellowhammer song but lacks drawn-out final note. Call a high-pitched 'tic'.

Habitat Breeds on farmland and other open country in southern Europe. In Britain only in parts of south Devon and Cornwall, on coastal farmland with mature hedgerows, weedy set-aside and stubble fields.

Food and habits Insectivorous when breeding, otherwise seed-eater. Feeds mainly on ground. Social in winter, flocks with other seed-eaters.

Reed Bunting

Emberiza schoeniclus

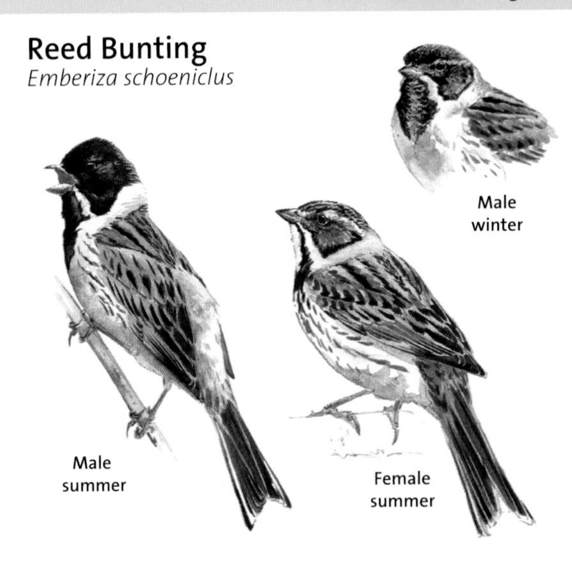

Size and description 15cm. Summer male has a rich brown back streaked darker, grey-brown rump, blackish tail with white outer feathers and whitish upperparts. Crown and face are black, with a white collar running into white moustachial streaks; throat and upper breast also black. Winter male, female and juvenile less boldly marked.

Voice Call 'tsee-you'. Song a repetitive 'tsit tsit tsrit tsrelitt'.

Habitat Resident in Britain and western Europe. Breeds mainly on marshland, both inland and coastal. More dispersed in winter, visiting gardens and farmland.

Food and habits Feeds mainly on seeds. Often perches on reed stems or telegraph wires. Nest a grassy cup concealed low in vegetation.

Index

English Names

Auk, Little 139
Avocet 82
Bittern 61
Bunting, Cirl 188
Corn 187
Lapland 185
Reed 189
Snow 186
Buzzard, Common 69
Rough-legged 70
Chiffchaff, Common 171
Chough 176
Coot 79
Cormorant 59
Crake, Spotted 77
Crow, Carrion 178
Hooded 179
Cuckoo 142
Curlew 102
Diver, Black-throated 43
Great Northern 44
Red-throated 42
Dove, Rock 141
Duck, Long-tailed 35
Tufted 30
Dunlin 95
Eagle, White-tailed 66
Egret, Great White 63
Little 62
Eider, Common 32
King 33
Falcon, Peregrine 75
Flycatcher, Pied 174
Fulmar 50
Gadwall 23
Gannet 58
Garganey 27
Godwit, Bar-tailed 100
Black-tailed 99
Goldeneye 38
Goldfinch 182
Goosander 41
Goose, Barnacle 20
Bean 15
Brent 18
Canada 18
Greylag 17
Pink-footed 15
White-fronted 16
Grebe, Black-necked 49
Great Crested 46
Little 45
Red-necked 47
Slavonian 48
Greenshank 105
Guillemot, Black 138
Common 136
Gull, Black-headed 129
Caspian 123
Common 120
Glaucous 126
Great Black-backed 127
Herring 124
Iceland 125
Lesser Black-backed 121
Little 117
Mediterranean 116
Ring-billed 119
Sabine's 118
Yellow-legged 122
Harrier, Hen 68
Marsh 67
Heron, Grey 64
Hobby 74
Jackdaw 177
Kestrel 72
Kingfisher 147
Kittiwake 128
Knot 88
Lapwing 87
Lark, Shore 151
Linnet 183
Mallard 25
Martin, House 154
Sand 152
Merganser, Red-breasted 40
Merlin 73
Moorhen 78
Osprey 71
Ouzel, Ring 165
Owl, Barn 143
Long-eared 144
Short-eared 145
Oystercatcher 80
Petrel, European Storm 56–7
Leach's 56–7
Phalarope, Grey 111
Red-necked 110
Pigeon, Feral 141
Pintail 26
Pipit, Meadow 155
Rock 156
Water 157
Plover, Golden 85
Grey 86
Little Ringed 83
Ringed 84
Pochard 29
Puffin 140
Rail, Water 76
Raven 180
Razorbill 137
Redshank 104
Spotted 103
Redstart, Black 161
Ruff 96
Sanderling 89
Sandpiper, Common 108
Curlew 93
Green 106
Pectoral 92

Purple 94
Wood 107
Scaup 31
Scoter, Common 35
Surf 36
Velvet 37
Skua, Arctic 113
Great 115
Long-tailed 114
Pomarine 112
Shearwater, Balearic 55
Cory's 51
Great 54
Manx 52–3
Sooty 52–3
Shelduck 21
Shoveler 28
Skylark 150
Smew 39
Snipe, Common 98
Jack 97
Spoonbill 65
Starling 181
Stilt, Black-winged 81
Stint, Little 90
Temminck's 91
Stonechat 163
Swallow 153
Swan, Bewick's 12–13
Mute 14
Whooper 12–13
Swift, Common 146
Teal, Common 24
Tern, Arctic 135
Black 131
Common 133
Little 130
Roseate 134
Sandwich 132
Tit, Bearded 175
Turnstone 109
Twite 184
Wagtail, Blue-headed 158
Grey 159
Pied 160
White 160
Yellow 158
Warbler, Cetti's 166
Grasshopper 167
Sedge 168
Reed 169
Willow 162
Yellow-browed 173
Wheatear 164
Whimbrel 101
Whitethroat, Common 170
Wigeon 22
Woodpecker, Green 149
Wryneck 148

Scientific names

Acrocephalus schoenobaenus 168
scirpaceus 169
Actitis hypoleucos 108
Alauda arvensis 150
Alca torda 137
Alcedo atthis 147
Alle alle 139
Anas acuta 26
clypeata 28
crecca 24
penelope 22
platyrhynchos 25
querquedula 27
strepera 23
Anser albifrons 16
anser 16
brachyrhynchus 15
fabalis 15
Anthus petrosus 156
pratensis 155
spinoletta 157
Apus apus 146
Ardea alba 63
cinerea 64
Arenaria interpres 109
Asio flammeus 145
otus 144
Aythya ferina 29
fuligula 30
marila 31
Botaurus stellaris 61
Branta bernicla 18
canadensis 18
leucopsis 20
Bucephala clangula 38
Buteo buteo 69
lagopus 70
Calcarius lapponicus 185
Calidris alba 89
alpina 95
canutus 88
ferruginea 93
maritima 94
melanotos 92
minuta 90
temminckii 91
Calonectris diomedea 51
Carduelis cannabina 183
carduelis 182
flavirostris 184
Cepphus grylle 138
Cettia cetti 166
Charadrius dubius 83
hiaticulus 84
Chlidonias niger 131
Chroicocephalus ridibundus 129
Circus aeruginosus 67
cyaneus 68
Clangula hyemalis 35
Columba livia 141
var. *domestica* 141

Corvus corax 180
cornix 179
corone 178
monedula 177
Cuculus canorus 142
Cygnus columbianus 12
cygnus 12
olor 14
Delichon urbicum 154
Egretta garzetta 62
Emberiza calandra 187
cirlus 188
schoeniclus 189
Eremophila alpestris 151
Falco columbarius 73
peregrinus 75
subbuteo 74
tinnunculus 72
Ficedula hypoleuca 174
Fratercula arctica 140
Fulica atra 79
Fulmarus glacialis 50
Gallinago gallinago 98
Gallinula chloropus 78
Gavia arctica 43
immer 44
stellata 42
Haematopus ostralegus 80
Haliaetus albicilla 66
Himantopus himantopus 81
Hirundo rustica 153
Hydrobates pelagicus 56–7
Hydrocoleus minutus 117
Jynx torquilla 148
Larus argentatus 124
cachinnans 123
canus 120
delawarensis 119
fuscus 121
glaucoides 125
hyperboreus 126
marinus 127
melanocephalus 116
michahellis 122
Limosa lapponica 100
limosa 99
Locustella naevia 167
Lymnocryptes minimus 97
Melanitta fusca 37
nigra 35
perspicillata 36
Mergellus albellus 39
Mergus merganser 41
serrator 40
Morus bassanus 58
Motacilla alba 160
cinerea 159
flava 158
Numenius arquata 102
phaeopus 101
Oceanodroma leucorhoa 56–7
Oenanthe oenanthe 164
Pandion haliaetus 71
Panurus biarmicus 175
Phalacrocorax aristotelis 60
carbo 59
Phalaropus fulicarius 111
lobatus 110
Philomachus pugnax 96
Phoenicurus ochruros 161
phoenicurus 162
Phylloscopus collybita 171
inornatus 173
trochilus 172
Picus viridis 149
Platalea leucorodia 65
Plectrophenax nivalis 186
Pluvialis apricaria 85
squatarola 86
Podiceps auritus 48
cristatus 46
grisegena 47
nigricollis 49
Porzana porzana 77
Puffinus gravis 54
griseus 52–3
mauretanicus 55
puffinus 52–3
Pyrrhocorax pyrrhocorax 176
Rallus aquaticus 76
Recurvirostra avosetta 82
Riparia riparia 152
Rissa tridactyla 128
Saxicola torquatus 163
Somateria mollissima 32
spectabilis 33
Stercorarius longicaudus 114
parasiticus 113
pomarinus 112
skua 115
Sterna dougallii 134
hirundo 133
paradisea 135
sandvicensis 132
Sternula albifrons 130
Sturnus vulgaris 181
Sylvia communis 170
Tachybaptus ruficollis 45
Tadorna tadorna 21
Tringa erythropus 103
glareola 107
nebularia 105
ochropus 106
totanus 104
Turdus torquatus 165
Tyto alba 143
Uria aalge 136
Vanellus vanellus 87
Xema sabini 118